McDougal Littell

Algebra 1
Concepts and Skills

Larson Boswell Kanold Stiff

CHAPTER 5 # Resource Book

The Resource Book contains the wide variety of black-line masters available for Chapter 5. The blacklines are organized by lesson. Included are support materials for the teacher as well as practice, activities, applications, and assessment resources.

McDougal Littell
A HOUGHTON MIFFLIN COMPANY
Evanston, Illinois • Boston • Dallas

Contributing Authors

The authors wish to thank the following individuals for their contributions to the Chapter 5 Resource Book.

Rita Browning
Linda E. Byrom
José Castro
Rebecca S. Glus
Christine A. Hoover
Carolyn Huzinec
Karen Ostaffe
Jessica Pflueger
Barbara L. Power
James G. Rutkowski
Michelle Strager

ISBN: 0-618-07855-X

456789-DWI-04 03 02

Contents

Contents

Contents

Descriptions of Resources

This Chapter Resource Book is organized by lessons within the chapter in order to make your planning easier. The following materials are provided:

Tips for New Teachers These teaching notes provide both new and experienced teachers with useful teaching tips for each lesson, including tips about common errors and inclusion.

Parent Guide for Student Success This guide helps parents contribute to student success by providing an overview of the chapter along with questions and activities for parents and students to work on together.

Prerequisite Skills Review Worked-out examples are provided to review the prerequisite skills highlighted on the Study Guide page at the beginning of the chapter. Additional practice is included with each worked-out example.

Strategies for Reading Mathematics The first page teaches reading strategies to be applied to the current chapter and to later chapters. The second page is a visual glossary of key vocabulary.

Lesson Plans and Lesson Plans for Block Scheduling This planning template helps teachers select the materials they will use to teach each lesson from among the variety of materials available for the lesson. The block-scheduling version provides additional information about pacing.

Warm-Up Exercises and Daily Homework Quiz The warm-ups cover prerequisite skills that help prepare students for a given lesson. The quiz assesses students on the content of the previous lesson. (Transparencies also available)

Activity Support Masters These blackline masters make it easier for students to record their work on selected activities in the Student Edition.

Alternative Lesson Openers An engaging alternative for starting each lesson is provided from among these four types: *Application, Activity, Graphing Calculator,* or *Visual Approach.* (Color transparencies also available)

Graphing Calculator Activities with Keystrokes Keystrokes for four models of calculators are provided for each Technology Activity in the Student Edition, along with alternative Graphing Calculator Activities to begin selected lessons.

Practice A and B These exercises offer additional practice for the material in each lesson, including application problems. There are two levels of practice for each lesson: A (transitional) and B (average).

Contents

Reteaching with Practice These two pages provide additional instruction, worked-out examples, and practice exercises covering the key concepts and vocabulary in each lesson.

Quick Catch-Up for Absent Students This handy form makes it easy for teachers to let students who have been absent know what to do for homework and which activities or examples were covered in class.

Learning Activities These enrichment activities apply the math taught in the lesson in an interesting way that lends itself to group work.

Interdisciplinary Applications/Real-Life Applications Students apply the mathematics covered in each lesson to solve an interesting interdisciplinary or real-life problem.

Challenge: Skills and Applications Teachers can use these exercises to enrich or extend each lesson.

Quizzes The quizzes can be used to assess student progress on two or three lessons.

Chapter Review Games and Activities This worksheet offers fun practice at the end of the chapter and provides an alternative way to review the chapter content in preparation for the Chapter Test.

Chapter Tests A and B These are tests that cover the most important skills taught in the chapter. There are two levels of test: A (transitional) and B (average).

SAT/ACT Chapter Test This test also covers the most important skills taught in the chapter, but questions are in multiple-choice and quantitative-comparison format. (See *Alternative Assessment* for multi-step problems.)

Alternative Assessment with Rubrics and Math Journal A journal exercise has students write about the mathematics in the chapter. A multi-step problem has students apply a variety of skills from the chapter and explain their reasoning. Solutions and a 4-point rubric are included.

Project with Rubric The project allows students to delve more deeply into a problem that applies the mathematics of the chapter. Teacher's notes and a 4-point rubric are included.

Cumulative Review These practice pages help students maintain skills from the current chapter and preceding chapters.

Tips for New Teachers

For use with Chapter 5

LESSON 5.1

TEACHING TIP Many examples for real-life situations can be solved without making a graph. Ask your students to name some advantages of making a graph. For example, using the graph it is easy to tell whether the linear model is increasing or decreasing. In addition, the graph can be used to quickly estimate other values of the linear function. You can use this discussion when introducing Lesson 5.5, Modeling with Linear Equations.

LESSON 5.2

COMMON ERROR Some students may have trouble working with the point-slope form of a linear equation. They might plug in values for x instead of x_1 or for y instead of y_1. Tell your students that a letter with a subscript, such as x_1 or y_1, represents a specific point and its coordinate values must be substituted into the equation. The final equation should contain the variables x and y.

COMMON ERROR Students might forget to account for the minus signs in the point-slope formula or reverse the x and y coordinates as (y_1, x_1).

LESSON 5.3

TEACHING TIP Ask your students if two points are enough information to graph a line. Since the answer is yes, it must also be possible to find the equation of a line given two points on it.

TEACHING TIP After finding the slope of the line, some students wonder which point they should use to find the y-intercept b. To show them that it does not matter, give two points on a line. Have one half of your class find the equation of the line by using the first point to find b and have the other half use the second point. They should all get the same answer.

TEACHING TIP Include some class examples where the two given points are on a horizontal or a vertical line. Once the students find the corresponding slope, whether zero or undefined, they should realize that these are special lines. Review how to write and graph the equations for vertical and horizontal lines.

LESSON 5.4

COMMON ERROR Some students believe that the different forms of a linear equation represent different lines. Tell your students that the same line can be written in several different forms. The forms of a linear equation can be compared to the representations of a number. For instance, the number five can be represented by spelling it out as "five," by the numeral "5," by the Roman numeral "V," or even by drawing five dots as in a die. These are all just different representations of the same thing.

TEACHING TIP Ask students what technique they would use to graph a linear equation based on the form in which it was given to them. Show them how finding the intercepts is an easy method when the equation is in standard form. This will help to review the graphing methods covered in Chapter 4 and will show students that they do not always have to find the slope-intercept form of a line to graph it.

CHAPTER 5 CONTINUED

CHAPTER

5

CONTINUED

Chapter Support

Tips for New Teachers
For use with Chapter 5

LESSON 5.5

TEACHING TIP Students may have difficulty assigning values to the variables. Emphasize that the slope is the rate of change. For example in Example 1 on page 298, the number of theaters increased by 750 each year. In addition, assigning x and y values can be callenging. If the students were drawing a line or bar graph, the time is horizontal and the quantity is vertical. When the Cartesian coordinate is drawn, remind the students that the time and the quantity should be the same as a graph. Thus, the time is the x-value and the quantity is the y-value (t_1, y_1).

LESSON 5.6

TEACHING TIP Students may need help getting started with the Checkpoint on page 308. To help them find the slope, locate the points $(10, -2)$ and $(7, -3)$. On the graph (with their finger), they're traveling 1 unit down (negative) and 3 units left (negative) which yields a positive slope of $\frac{1}{3}$. Ask the class what they think the slope of the path of the ship would be (-3). Using slope-intercept form, substitute the y-value, the perpendicular slope, and the x-value to find b: $4 = (-3)(4) + b$. This will yield $y = -3x + 16$.

INCLUSION You can help students with limited English proficiency with the vocabulary by reviewing perpendicularity. Additionally, some students may have difficulty with the product of the slopes being -1. Some review of fractions and multiplicative identity should help. Using an example of a slope of 1 and -1 may confuse students into thinking that finding perpendicularity only requires a negative value. Beginning the lesson with fractions and their reciprocals will enable them to better understand the concept.

Outside Resources

BOOKS/PERIODICALS

VanDyke, Frances. "Relating to Graphs in Introductory Algebra." *Mathematics Teacher* (September 1994); pp. 427–432, 438, 439.

ACTIVITIES/MANIPULATIVES

Anderson, Edwin D. and Jim Nelson. "An Introduction to the Concept of Slope." *Mathematics Teacher* (January 1994); pp. 27–30, 37–41.

SOFTWARE

Dugdale, Sharon and David Kibbey. *Green Globs and Graphing Equations*. Introductory graphing concepts, tutorials, exploring graphs. Pleasantville, NY; Sunburst Communications.

VIDEOS

Algebra in Simplest Terms. Linear equations. Burlington, VT; Annenburg/CPB Collection, 1991.

NAME _____ DATE _____

Parent Guide for Student Success

For use with Chapter 5

Chapter Overview One way that you can help your student succeed in Chapter 5 is by discussing the lesson goals in the chart below. When a lesson is completed, ask your student to interpret the lesson goals for you and to explain how the mathematics of the lesson relates to one of the key applications listed in the chart.

Lesson Title	Lesson Goals	Key Applications
5.1 Slope-Intercept Form	Use slope-intercept form to write the equation of a line.	• Space Shuttle • Olympics • Old Faithful
5.2 Point-Slope Form	Use point-slope form to write an equation of a line.	• Scuba Diving • Travel
5.3 Writing Linear Equations Given Two Points	Write an equation of a line given two points on the line.	• Airplanes • Chunnel • Speed of Sound
5.4 Standard Form	Write an equation of a line in standard form.	• Bird Seed Mixture
5.5 Modeling with Linear Equations	Write and use a linear equation to solve a real-life problem.	• Movie Theaters • Mountain Climbing • Car Costs
5.6 Perpendicular Lines	Write equations of perpendicular lines.	• Helicopters • Construction

Study Tip

Create a Practice Test is the study tip featured in Chapter 5 (see page 268). Encourage your student to create a test, exchange tests with a classmate, score each other's tests, and plan further study. You may wish to discuss with your student the kinds of test questions the teacher is likely to ask and to help your student in developing a study plan.

NAME _____ DATE _____

Parent Guide for Student Success

For use with Chapter 5

Key Ideas Your student can demonstrate understanding of key concepts by working through the following exercises with you.

Lesson	Exercise
5.1	Write the equation of the line whose slope is $-\frac{1}{2}$ and whose y intercept is -3.
5.2	Write in point-slope form an equation of the line that is parallel to $y = \frac{1}{2}x - 9$ and passes through the point $(-4, 7)$.
5.3	Write an equation in slope-intercept form of the line that passes through the points $(-2, 9)$ and $(6, -7)$.
5.4	Write an equation in standard form of the line that passes through the points $(-5, -4)$ and $(-2, 8)$.
5.5	You have \$60 to spend at a sale where sweatshirts are \$10 each and blue jeans are \$15 each. Your parents agree to pay the tax. Write an equation that models the different number of sweatshirts and blue jeans you can buy.
5.6	Write in slope-intercept form the equation of the line that passes through $(-1, 3)$ and is perpendicular to $y = \frac{1}{2}x - 7$.

Home Involvement Activity

You will need: A calculator

Directions: Find the amount your family spent on utilities (or one specific utility) last year and at least 3 years ago. Use the data to find a linear model for the amount y spent t years after 1990. Use the model to estimate how much you can expect to spend on utilities next year.

Answers

5.1: $y = -\frac{1}{2}x - 3$ **5.2:** $y - 7 = \frac{1}{2}(x + 4)$ **5.3:** $y = -2x + 5$ **5.4:** $y - 4x = 16$
5.5: $10x + 15y = 60$ where x is the number of sweatshirts and y is the number of blue jeans **5.6:** $y = -2x + 1$

NAME _____ DATE _____

Prerequisite Skills Review

For use before Chapter 5

EXAMPLE 1 *Solving Equations With Variables on Both Sides*

Solve the equation.

$8(x - 1) = 4x - 17 + x$

SOLUTION

$8(x - 1) = 4x - 17 + x$	Write the original equation.
$8x - 8 = 4x - 17 + x$	Use the distributive property.
$8x - 8 = 5x - 17$	Add like terms.
$3x - 8 = -17$	Subtract $5x$ from each side.
$3x = -9$	Add 8 to each side.
$x = -3$	Divide each side by 3.

Exercises for Example 1

Solve the equation. If there is no solution, write *no solution*.

1. $2(1 - x) + 3x = -4(x + 2)$ **2.** $12x - 8 = 3(4x + 11)$

3. $\frac{1}{5}(25x + 60) = 33 - 4(x - 6)$ **4.** $-\frac{2}{9}(18x - 9) = 6\left(x - \frac{1}{2}\right)$

EXAMPLE 2 *Plotting Points in a Coordinate Plane*

Plot the ordered pairs in a coordinate plane.

$A(6, -1), B(-3, -7), C(0, 2)$

SOLUTION

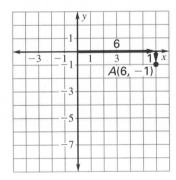 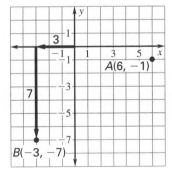 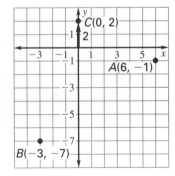

To plot the point $(6, -1)$ start at the origin. Move 6 units to the right and 1 unit down.

To plot the point $(-3, -7)$ start at the origin. Move 3 units to the left and 7 units down.

To plot the point $(0, 2)$ start at the origin. Move 2 units up.

Algebra 1
Chapter 5 Resource Book

NAME _____ DATE _____

Prerequisite Skills Review

For use before Chapter 5

Exercises for Example 2

Plot the ordered pairs in a coordinate plane.

5. $A(5, 2), B(-3, 8), C(7, -2)$

6. $A(-4.1, -3), B(-1, -1), C(2.1, 4)$

7. $A(-1, 0), B(0, -6), C(\frac{1}{2}, -4)$

8. $A(-3, 0), B(1.5, 3), C(5, -3.5)$

EXAMPLE 3 *Using Intercepts to Graph Equations*

Find the x-intercept and the y-intercept of the equation. Graph the line.

$5x + 8y = 24$

SOLUTION

Find the intercepts by substituting 0 for y and then 0 for x.

$5x + 8y = 24$ $5x + 8y = 24$

$5x + 8(0) = 24$ $5(0) + 8y = 24$

$5x = 24$ $8y = 24$

$x = \frac{24}{5}$ $y = 3$

The x-intercept is $\frac{24}{5} = 4\frac{4}{5}$. The y-intercept is 3.

Draw a line that includes the points $\left(\frac{24}{5}, 0\right)$ and $(0, 3)$.

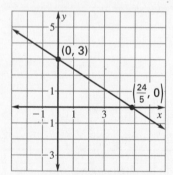

Exercises for Example 3

**Find the x-intercept and the y-intercept in the equation.
Graph the line.**

9. $5x + y = 25$

10. $-3x = 11y - 33$

11. $-x - 6y = 35 + 4x$

12. $x + 8y = 10$

Strategies for Reading Mathematics

For use with Chapter 5

Strategy: Translating Words into Symbols

When you use mathematics to solve real-life problems, you will often translate the words into symbols. Read the following problem.

> A family pays a one-time registration fee of $35 to enroll their child in daycare. The weekly charge thereafter is $200 per week. How much will the family pay for daycare?

The steps outlined below can help you interpret this problem and other problems like it.

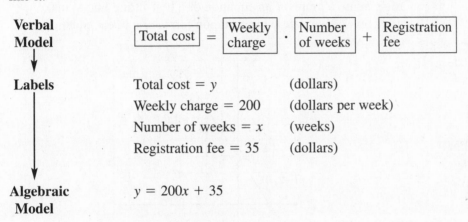

Verbal Model

$$\boxed{\text{Total cost}} = \boxed{\begin{array}{c}\text{Weekly} \\ \text{charge}\end{array}} \cdot \boxed{\begin{array}{c}\text{Number} \\ \text{of weeks}\end{array}} + \boxed{\begin{array}{c}\text{Registration} \\ \text{fee}\end{array}}$$

Labels

Total cost $= y$ (dollars)

Weekly charge $= 200$ (dollars per week)

Number of weeks $= x$ (weeks)

Registration fee $= 35$ (dollars)

Algebraic Model

$y = 200x + 35$

STUDY TIP

Use a Problem-Solving Model

To model a real-life situation using mathematics, begin by writing a verbal model based on the relationships in the situation. Next assign labels to the pieces of your model. Then use your labels to write an algebraic model for the situation.

Questions

1. How much does it cost for three weeks of daycare? for five weeks of daycare? What is the rate of change in dollars per week?

2. If you graphed the equation, what would be the *y*-intercept?

3. Suppose the registration fee was $40 and the weekly charge was $225. Write an equation to represent this situation.

4. Write a verbal model, labels, and an algebraic model for the following situation.

Rachel receives a weekly allowance of $3.00 for doing certain household chores. She can earn more money by doing additional chores for $1.50 per chore. How much money can Rachel earn in a week?

Strategies for Reading Mathematics

For use with Chapter 5

Visual Glossary

The Study Guide on page 268 lists the key vocabulary for Chapter 5. Use the visual glossary below to help you understand some of the key vocabulary in Chapter 5. You may want to copy these diagrams into your notebook and refer to them as you complete the chapter.

GLOSSARY

slope-intercept form (p. 269) A linear equation written in the form $y = mx + b$. The slope of the line is m. The y-intercept is b.

point-slope form (p. 278) An equation in the form $y - y_1 = m(x - x_1)$ of a line that passes through a given point (x_1, y_1) with a slope of m.

standard form (p. 291) A linear equation of the form $Ax + By = C$ where A and B are not both zero.

linear model (p. 298) A linear function that is used to model a real-life situation.

rate of change (p. 298) A comparison of two quantities that are changing.

perpendicular (p. 306) Two lines that intersect at a right angle.

Graphing Equations of a Line

You can use a graph or an equation of a best-fitting line to make predictions. There are three common forms of a linear equation.

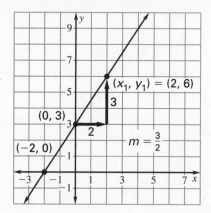

slope-intercept form	point-slope form	standard form
$y = mx + b$	$y - y_1 = m(x - x_1)$	$Ax + By = C$
$y = \frac{3}{2}x + 3$	$y - 6 = \frac{3}{2}(x - 2)$	$-3x + 2y = 6$

TEACHER'S NAME _____ CLASS _____ ROOM _____ DATE _____

Lesson Plan
1-day lesson (See *Pacing the Chapter,* TE page 266A)   **For use with pages 269–275**

GOAL **Use slope-intercept form to write an equation of a line.**

State/Local Objectives _____

_____

✓ **Check the items you wish to use for this lesson.**

STARTING OPTIONS
____ Prerequisite Skills Review: CRB pages 5–6
____ Strategies for Reading Mathematics: CRB pages 7–8
____ Warm-Up: CRB page 11 or Transparencies

TEACHING OPTIONS
____ Lesson Opener: CRB page 12 or Transparencies
____ Examples 1–3: SE pages 269–271
____ Extra Examples: TE pages 270–271 or Transparencies
____ Checkpoint Exercises: SE pages 269–271
____ Concept Check: TE page 271
____ Guided Practice Exercises: SE page 272

APPLY/HOMEWORK
Homework Assignment
____ Transitional: pp. 272–275, Exs. 13–15, 22, 23, 28, 29, 34–43, 52–55, 57–83 odd
____ Average: pp. 272–275, Exs. 16–18, 24, 25, 30, 31, 44–46, 52–55, 56–82 even
____ Advanced: pp. 272–275, Exs. 19–21, 26, 27, 32, 33, 47–55*, 59–61, 65–67, 71–73, 79–83; EC: CRB p. 19

Reteaching the Lesson
____ Practice Masters: CRB pages 13–14 (Level A, Level B)
____ Reteaching with Practice: CRB pages 15–16 or Practice Workbook with Examples;
Resources in Spanish
____ Personal Student Tutor: CD-ROM

Extending the Lesson
____ Interdisciplinary/Real-Life Applications: CRB page 18
____ Challenge: CRB page 19

ASSESSMENT OPTIONS
____ Daily Quiz (5.1): TE page 275, CRB page 22, or Transparencies
____ Standardized Test Practice: SE page 275; STP Workbook; Transparencies

Notes _____

_____

_____

Lesson Plan for Block Scheduling

Half-block lesson (See *Pacing the Chapter,* TE page 266A)   **For use with pages 269–275**

GOAL **Use slope-intercept form to write an equation of a line.**

State/Local Objectives _____

_____

_____

✓ Check the items you wish to use for this lesson.

STARTING OPTIONS

____ Prerequisite Skills Review: CRB pages 5–6
____ Strategies for Reading Mathematics: CRB pages 7–8
____ Warm-Up: CRB page 11 or Transparencies

TEACHING OPTIONS

____ Lesson Opener: CRB page 12 or Transparencies
____ Examples 1–3: SE pages 269–271
____ Extra Examples: TE pages 270–271 or Transparencies
____ Checkpoint Exercises: SE pages 269–271
____ Concept Check: TE page 271
____ Guided Practice Exercises: SE page 272

APPLY/HOMEWORK

Homework Assignment (See also the assignment for Lesson 5.2.)
____ Block Schedule: pp. 272–275: Exs. 16–18, 24, 25, 30, 31, 44–46, 54, 55, 56–82 even

Reteaching the Lesson
____ Practice Masters: CRB pages 13–14 (Level A, Level B)
____ Reteaching with Practice: CRB pages 15–16 or Practice Workbook with Examples;
Resources in Spanish
____ Personal Student Tutor: CD-ROM

Extending the Lesson
____ Interdisciplinary/Real-Life Applications: CRB page 18
____ Challenge: CRB page 19

ASSESSMENT OPTIONS

____ Daily Quiz (5.1): TE page 275, CRB page 22, or Transparencies
____ Standardized Test Practice: SE page 275; STP Workbook; Transparencies

Notes _____

_____

_____

CHAPTER PACING GUIDE	
Day	**Lesson**
1	**5.1 (all)**; 5.2 (begin)
2	5.2 (end); 5.3 (all)
3	5.4 (all); 5.5 (all)
4	5.6 (all)
5	Ch. 5 Review and Assess

Algebra 1
Chapter 5 Resource Book

LESSON

5.1

NAME _____ DATE _____

WARM-UP EXERCISES

For use before Lesson 5.1, pages 269–275

Simplify each expression.

1. $\dfrac{2 - 3}{-4 + 1}$

2. $\dfrac{-2 - 1}{2 + (-3)}$

Evaluate for $x = -1$, 0, and 2.

3. $f(x) = 3x - 7$

4. $f(x) = 205 + 32x$

DAILY HOMEWORK QUIZ

For use after Lesson 4.8, pages 252–258

1. Decide whether the relation is a function. If it is a function, give the domain and range.

Input	0	4	8	12	16
Output	1	2	3	4	5

2. Decide whether the graph represents a function. Explain your reasoning.

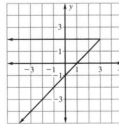

3. Evaluate $f(x) = 3x - 2$ when $x = -4$.

4. You ride your bike at a speed of 12 miles per hour. Write a linear function that models your distance d as a function of your time t in hours.

Algebra 1
Chapter 5 Resource Book

1. a. Graph the equation $y = 2x + 1$.

 b. What is the *y*-intercept of the line? You may want to use the trace feature or table feature to help you answer this question.

 c. What is the slope of the line? You may want to use the table feature to help you answer this question.

 d. Compare your answers for parts (b) and (c) to the equation in part (a). What do you notice?

2. a. Graph the equation $y = x - 3$.

 b. What is the *y*-intercept of the line? You may want to use the trace feature or table feature to help you answer this question.

 c. What is the slope of the line? You may want to use the table feature to help you answer this question.

 d. Compare your answers for parts (b) and (c) to the equation in part (a). What do you notice?

3. a. Graph the equation $y = -\dfrac{1}{2}x + 2$

 b. What is the *y*-intercept of the line? You may want to use the trace feature or table feature to help you answer this question.

 c. What is the slope of the line? You may want to use the table feature to help you answer this question.

 d. Compare your answers for parts (b) and (c) to the equation in part (a). What do you notice?

4. Consider the equation $y = 5x - 3$. Use your answers to Exercises 1–3 to predict the slope and *y*-intercept of this equation. Then graph the equation to check your prediction.

NAME _____ DATE _____

Practice A
For use with pages 269–275

Find the slope and *y*-intercept of the line.

1. $y = 2x + 5$ **2.** $y = -4x + 1$ **3.** $y = x - 5$

4. $y = \frac{1}{2}x$ **5.** $y = 3 + 2x$ **6.** $2y = 4x - 3$

Write in slope-intercept form the equation of the line.

7. The slope is 1; the *y*-intercept is 0.   **8.** The slope is -2; the *y*-intercept is 4.

9. The slope is -3; the *y*-intercept is -5.   **10.** The slope is 6; the *y*-intercept is -1.

11. The slope is 0; the *y*-intercept is 9.   **12.** The slope is -6; the *y*-intercept is -2.

13. The slope is 2; the *y*-intercept is -8.   **14.** The slope is -4; the *y*-intercept is 11.

15. The slope is 5; the *y*-intercept is 5.   **16.** The slope is -5; the *y*-intercept is -4.

17. The slope is $-\frac{3}{5}$; the *y*-intercept is 3.   **18.** The slope is $\frac{8}{9}$; the *y*-intercept is $-\frac{1}{2}$.

Write in slope-intercept form the equation of the line shown in the graph.

19.
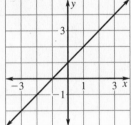

20.
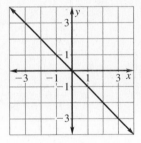

21.
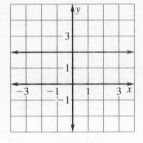

22.
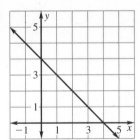

23.
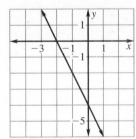

24.
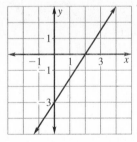

Algebra 1
Chapter 5 Resource Book

NAME _____ DATE _____

Practice B

For use with pages 269–275

Write in slope-intercept form the equation of the line.

1. The slope is 2; the y-intercept is 3.

2. The slope is 5; the y-intercept is 0.

3. The slope is 4; the y-intercept is -3.

4. The slope is -5; the y-intercept is 1.

5. The slope is -3; the y-intercept is -2.

6. The slope is 0; the y-intercept is -5.

7. The slope is $\frac{1}{2}$; the y-intercept is -8.

8. The slope is $-\frac{3}{4}$; the y-intercept is 9.

9. The slope is $-\frac{1}{5}$; the y-intercept is 3.

10. The slope is $\frac{4}{5}$; the y-intercept is -7.

11. The slope is $\frac{1}{3}$; the y-intercept is $\frac{2}{3}$.

12. The slope is $-\frac{4}{3}$; the y-intercept is $\frac{7}{8}$.

Write in slope-intercept form the equation of the line shown in the graph.

13.

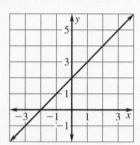

14.

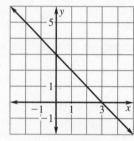

15.

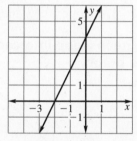

16.

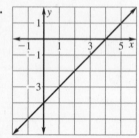

17.

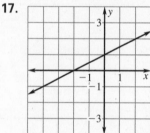

18.

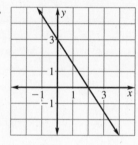

Algebra 1
Chapter 5 Resource Book

NAME _____ DATE _____

Reteaching with Practice

For use with pages 269–275

GOAL **Use slope-intercept form to write an equation of a line.**

> **VOCABULARY**
>
> The **slope-intercept form** of the equation of a line with slope m and y-intercept b is $y = mx + b$.

EXAMPLE 1 *Equation of a Line*

Write an equation of the line with slope 4 and y-intercept -3.

SOLUTION

$$y = mx + b \qquad \text{Write slope-intercept form.}$$

$$y = 4x + (-3) \qquad \text{Substitute 4 for } m \text{ and } -3 \text{ for } b.$$

$$y = 4x - 3 \qquad \text{Simplify.}$$

Exercises for Example 1

Write an equation of the line in slope-intercept form.

1. The slope is -2; the y-intercept is 5.   **2.** The slope is 1; the y-intercept is -4.

3. The slope is 0; the y-intercept is 2.   **4.** The slope is 3; the y-intercept is 6.

EXAMPLE 2 *Using a Graph to Write an Equation*

Write an equation of the line shown using slope-intercept form.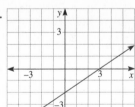

SOLUTION

Write the slope-intercept form $y = mx + b$.
Find the slope of the line. Let $(0, -2)$ be (x_1, y_1)
and $(3, 0)$ be (x_2, y_2).

$$m = \frac{\text{rise}}{\text{run}} = \frac{y_2 - y_1}{x_2 - x_1} = \frac{0 - (-2)}{3 - 0} = \frac{2}{3}$$

Use the graph to find the y-intercept b. The y-intercept is -2.

Substitute $\frac{2}{3}$ for m and -2 for b in $y = mx + b$: $y = \frac{2}{3}x - 2$.

Exercises for Example 2

Write an equation of the line in slope-intercept form.

5. 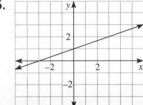   **6.**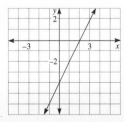

NAME _____ DATE _____

Reteaching with Practice

For use with pages 269–275

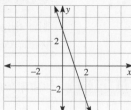

EXAMPLE 3 *Modeling Negative Slope*

Write an equation of the line shown using slope-intercept form.

SOLUTION

Write the slope-intercept form $y = mx + b$.
Find the slope of the line. Let $(0, 3)$ be (x_1, y_1)
and $(1, 0)$ be (x_2, y_2).

$$m = \frac{\text{rise}}{\text{run}} = \frac{y_2 - y_1}{x_2 - x_1} = \frac{0 - 3}{1 - 0} = \frac{-3}{1} = -3$$

Use the graph to find the y-intercept b. The y-intercept is 3.

Substitute -3 for m and 3 for b in $y = mx + b$.

$$y = -3x + 3$$

Exercises for Example 3

Write an equation of the line in slope-intercept form.

7.

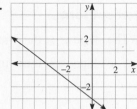

8.

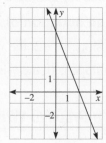

Algebra 1
Chapter 5 Resource Book

NAME _____ DATE _____

Quick Catch-Up for Absent Students

For use with pages 269–275

The items checked below were covered in class on (date missed) _____

Lesson 5.1: Slope-Intercept Form

_____ **Goal:** Use slope-intercept form to write an equation of a line.

Material Covered:

_____ Example 1: Equation of a Line

_____ Student Help: Study Tip

_____ Example 2: Use a Graph to Write an Equation

_____ Student Help: Study Tip

_____ Example 3: Model Negative Slope

Vocabulary:

slope-intercept form, p. 269

_____ Other (specify) _____

_____

Homework and Additional Learning Support

_____ Textbook (specify) _pp. 272–275_____

_____ *Reteaching with Practice* worksheet (specify exercises)_____

_____ *Personal Student Tutor* for Lesson 5.1

NAME _____ DATE _____

Interdisciplinary Application

For use with pages 269–275

Break-Even Analysis

ECONOMICS   One of the most important aspects of business management
is determining the price a company will charge for goods and services.
Companies will often choose to price products to maximize profits. Another
consideration is the cost associated with developing the good or service.
One tool used by managers to analyze costs is break-even analysis.

A break-even analysis will determine the quantity of a product that must be
sold before the seller begins to make a profit. The analysis takes into
consideration variable costs and fixed costs. Variable costs change with the
quantity of product produced while fixed costs remain constant. Examples
of fixed costs are rent, insurance, administrative salaries, and equipment.
Some variable costs are production worker's wages, material expense, and
utilities expense.

By graphing both a revenue line and a cost line, a company can then
determine a break-even point. This occurs when the revenue and cost lines
intersect. This intersection point will be the quantity needed to at least cover
costs of production. Any greater quantity will then start generating a profit.

In Exercises 1–3, use the following information.

A store can purchase T-shirts for $7 each. It has fixed costs of $2500.
Each T-shirt is sold for $18.

   1. Write a linear equation for both cost and revenue.

   2. Graph both the cost line and the revenue line.

   3. Estimate the break-even point.

In Exercises 4–6, use the following information.

A hotdog stand can purchase hotdogs for $.35 each and buns for $.20 each.
It has fixed costs of $50. Each hotdog is sold for $1.

   4. Write a linear equation for both cost and revenue.

   5. Graph both the cost and revenue lines.

   6. Estimate the break-even point.

Algebra 1
Chapter 5 Resource Book

NAME _____ DATE _____

Challenge: Skills and Applications

For use with pages 269–275

In Exercises 1–4, use the following information.

According to the census, the population of the United States was about 151 million people in 1950 and 249 million in 1990.

1. Find the slope of the line through the two points defined by the population data. What does this slope tell you?

2. Write a linear equation to model the population (in millions) of the United States t years after 1950.

3. Use the equation from Exercise 2 to find the population of the United States in 1970. According to the 1970 census, the population was about 203 million people. How close was the amount found with the model to the actual amount? Was this a good approximation? Explain.

4. Use the equation from Exercise 2 to predict the population of the United States in 2010.

In Exercises 5–7, use the following information.

In 1990, people in the United States spent about $285.7 billion on recreation. In 1995, they spent $402.5 billion.

5. Write a linear equation to model the amount (in billions of dollars) spent on recreation t years after 1990.

6. Use the equation from Exercise 5 to find the amount that people in the United States spent on recreation in 1996. The actual amount was $431.1 billion. How close was the amount found with the model to the actual amount? Do you think this is a good approximation? Explain.

7. Use the equation from Exercise 5 to predict the amount people in the United States will spend on recreation in 2000. Do you think the prediction is very accurate? Explain.

TEACHER'S NAME _____ CLASS _____ ROOM _____ DATE _____

Lesson Plan

2-day lesson (See *Pacing the Chapter,* TE page 266A)   **For use with pages 276–284**

GOAL **Use point-slope form to write the equation of a line.**

State/Local Objectives _____

_____

✓ Check the items you wish to use for this lesson.

STARTING OPTIONS
____ Homework Check (5.1): TE page 272; Answer Transparencies
____ Homework Quiz (5.1): TE page 275, CRB page 22, or Transparencies
____ Warm-Up: CRB page 22 or Transparencies

TEACHING OPTIONS
____ Developing Concepts: SE pages 276–277
____ Lesson Opener: CRB page 23 or Transparencies
____ Examples: Day 1: 1–3, SE pages 278–279; Day 2: 4, SE page 280)
____ Extra Examples: TE pages 279–280 or Transparencies; Internet Help at *www.mcdougallittell.com*
____ Checkpoint Exercises: Day 1: Exs. 1–5, SE pages 278–279; Day 2: Ex. 6, SE page 280
____ Concept Check: TE page 280
____ Guided Practice Exercises: SE page 281; Day 1: Exs. 1–10; Day 2: Exs. 11–13

APPLY/HOMEWORK
Homework Assignment
____ Transitional Day 1: pp. 281–284, Exs. 14, 15, 20–22, 26–28, 53–71 odd;
          Day 2: pp. 282–284, Exs. 35, 36, 41–43, 49–51, Quiz 1
____ Average Day 1: pp. 281–284, Exs. 16, 17, 23–25, 29–31, 52–70 even;
          Day 2: pp. 282–284, Exs. 37, 38, 41–45, 49–51, Quiz 1
____ Advanced Day 1: pp. 281–284, Exs. 18, 19, 32–34, 55–57, 61–63, 68–71;
          Day 2: pp. 282–284, Exs. 39, 40, 44–51*, Quiz 1; EC: CRB p. 30

Reteaching the Lesson
____ Practice Masters: CRB pages 24–25 (Level A, Level B)
____ Reteaching with Practice: CRB pages 26–27 or Practice Workbook with Examples;
          Resources in Spanish
____ Personal Student Tutor: CD-ROM

Extending the Lesson
____ Interdisciplinary/Real-Life Applications: CRB page 29
____ Challenge: CRB page 30

ASSESSMENT OPTIONS
____ Daily Quiz (5.2): TE page 283, CRB page 34, or Transparencies
____ Standardized Test Practice: SE page 283; STP Workbook; Transparencies
____ Quiz 5.1–5.2: SE page 284; CRB page 31; Resources in Spanish

Notes _____

_____

_____

Lesson 5.2

TEACHER'S NAME _____ CLASS _____ ROOM _____ DATE _____

Lesson Plan for Block Scheduling

1-block lesson (See *Pacing the Chapter,* TE page 266A)   For use with pages 276–284

GOAL   Use point-slope form to write the equation of a line.

State/Local Objectives _____

_____

_____

✓ Check the items you wish to use for this lesson.

CHAPTER PACING GUIDE	
Day	**Lesson**
1	5.1 (all); **5.2 (begin)**
2	**5.2 (end)**; 5.3 (all)
3	5.4 (all); 5.5 (all)
4	5.6 (all)
5	Ch. 5 Review and Assess

STARTING OPTIONS
____ Homework Check (5.1): TE page 272; Answer Transparencies
____ Homework Quiz (5.1): TE page 275,
  CRB page 22, or Transparencies
____ Warm-Up: CRB page 22 or Transparencies

TEACHING OPTIONS
____ Developing Concepts: SE pages 276–277
____ Lesson Opener: CRB page 23 or Transparencies
____ Examples: Day 1: 1–3, SE pages 278–279; Day 2: 4, SE page 280
____ Extra Examples: TE pages 279–280 or Transparencies; Internet Help at *www.mcdougallittell.com*
____ Checkpoint Exercises: Day 1: Exs. 1–5, SE pages 278–279; Day 2: Ex. 6, SE page 280)
____ Concept Check: TE page 280
____ Guided Practice Exercises: SE page 281; Day 1: Exs. 1–10; Day 2: Exs. 11–13

APPLY/HOMEWORK
Homework Assignment (See also the assignments for Lessons 5.1 and 5.3.)
____ Block Schedule: Day 1: pp. 281–284 Exs. 16, 17, 23–25, 29–31, 52–70 even;
  Day 2: pp. 281–284 Exs. 37, 38, 41–45, 49–51, Quiz 1

Reteaching the Lesson
____ Practice Masters: CRB pages 24–25 (Level A, Level B)
____ Reteaching with Practice: CRB pages 26–27 or Practice Workbook with Examples;
  Resources in Spanish
____ Personal Student Tutor: CD-ROM

Extending the Lesson
____ Interdisciplinary/Real-Life Applications: CRB page 29
____ Challenge: CRB page 30

ASSESSMENT OPTIONS
____ Daily Quiz (5.2): TE page 283, CRB page 34, or Transparencies
____ Standardized Test Practice: SE page 283; STP Workbook; Transparencies
____ Quiz 5.1–5.2: SE page 284; CRB page 31; Resources in Spanish

Notes _____

_____

_____

Lesson 5.2

LESSON

5.2

NAME _____ DATE _____

WARM-UP EXERCISES

For use before Lesson 5.2, pages 276–284

Available as
a transparency

Find the slope of the line through the points.

1. $(-2, 3), (5, -1)$ **2.** $(6, -2), (-1, -5)$

Solve each equation for y and then simplify.

3. $y - 21 = \dfrac{3}{5}(x + 10)$ **4.** $y + 1 = -2(x - 3)$

DAILY HOMEWORK QUIZ

For use after Lesson 5.1, pages 269–275

1. Write in slope-intercept form the equation of the line
with $m = -2$ and $b = 3$.

2. Identify the slope and
y-intercept of the line.

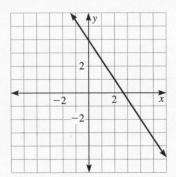

3. Write an equation in slope-
intercept form of the line
shown.

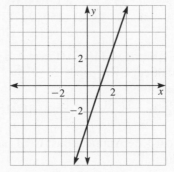

4. Identify the slope and the y-intercept of the line with
equation $y = \frac{1}{3}x - 5$.

5. The equation $y = -4x + 3$ represents your distance y,
in miles, from home after walking for x hours. Use the
equation to find your distance after walking for $\frac{1}{2}$ hour.

Lesson 5.2

LESSON

5.2

NAME _____ DATE _____

Activity Lesson Opener

For use with pages 276–284

Available as
a transparency

SET UP: Work with a partner.

YOU WILL NEED: • **straightedge** • **graph paper**

1. a. Graph the line that passes through the point $(1, 4)$ and has a slope of 2.

   b. On the same coordinate plane, graph the equation $y - 4 = 2(x - 1)$.

   c. What is true about the two lines?

   d. How is the ordered pair related to the equation in part (b)?

   e. How is the slope related to the equation in part (b)?

2. a. Graph the line that passes through the point $(-2, 3)$ and has a slope of -3.

   b. On the same coordinate plane, graph the equation $y - 3 = -3(x + 2)$.

   c. What is true about the two lines?

   d. How is the ordered pair related to the equation in part (b)?

   e. How is the slope related to the equation in part (b)?

3. a. Graph the line that passes through the point $(-3, -5)$ and has a slope of $\frac{2}{3}$.

   b. On the same coordinate plane, graph the equation $y + 5 = \frac{2}{3}(x + 3)$.

   c. What is true about the two lines?

   d. How is the ordered pair related to the equation in part (b)?

   e. How is the slope related to the equation in part (b)?

Lesson 5.2

NAME _____ DATE _____

Practice A
For use with pages 276–284

Write an equation in point-slope form of the line.

1.
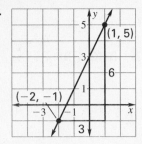

2.
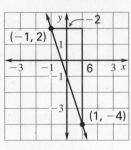

3.
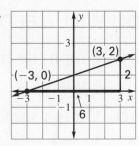

Write an equation in point-slope form of the line that passes through the given point and has the given slope.

4. $(2, 5), m = 3$

5. $(1, 4), m = 2$

6. $(-2, 0), m = \frac{1}{2}$

7. $(3, 7), m = 1$

8. $(-5, 8), m = -4$

9. $(0, -4), m = 9$

10. $(1, 1), m = 0$

11. $(-3, -4), m = -2$

12. $(6, -10), m = 5$

Write an equation in point-slope form of the line that is parallel to the given line and passes through the given point.

13. $y = 5x + 4, (0, 0)$

14. $y = 2x - 6, (2, 3)$

15. $y = 3x - 11, (9, 6)$

16. $y = -1x + 2, (8, -7)$

17. $y = 7x - 1, (1, -2)$

18. $y = \frac{1}{2}x + 8, (2, -7)$

19. $y = 3x - 9, (-5, -4)$

20. $y = -7x + 1, \left(-\frac{1}{2}, -\frac{1}{3}\right)$

21. $y = \frac{2}{3}x + 9, (4, -8)$

Classified Ads   **In Exercises 22 and 23, use the following information.**

It costs $2.00 per day to place a one-line ad in the classifieds plus a flat service fee. One day costs $3.00 and four days costs $9.00.

22. Write a linear equation in point-slope form that gives the cost in dollars, *y*, in terms of the number of days the ad appears, *x*.

23. Find the cost of a six-day ad.

Travel   **In Exercises 24 and 25, use the following information.**

You are driving from Grand Rapids, Michigan, to Detroit, Michigan. You leave Grand Rapids at 4:00 P.M. At 5:10 P.M. you pass through Lansing, Michigan, a distance of 65 miles.

24. Write a linear equation that gives the distance in miles, *d*, in terms of time, *t*. Let *t* represent the number of minutes since 4.00 P.M.

25. Approximately what time will you arrive in Detroit if it is 150 miles from Grand Rapids?

Lesson 5.2

NAME _____ DATE _____

Practice B

For use with pages 276–284

Write an equation in point-slope form of the line.

1.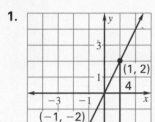

2.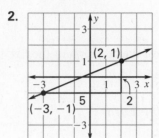

3.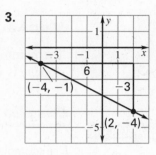

Write an equation in point-slope form of the line that passes through the given point and has the given slope.

4. $(-3, 24), m = -2$

5. $(-4, -2), m = -5$

6. $(0, -3), m = \frac{2}{3}$

7. $(6, -5), m = -4$

8. $(-7, 6), m = 0$

9. $(-3, -5), m = 6$

10. $(-12, 1), m = -6$

11. $(-14, 21), m = -\frac{1}{3}$

12. $(16, 4), m = -\frac{2}{3}$

Rewrite the equation in slope-intercept form.

13. $y - 2 = 1(x + 3)$

14. $y + 9 = 4(x - 3)$

15. $y - \frac{1}{2} = 2(x - 6)$

16. $y + 4 = 5(x + 2)$

17. $y - 3 = -2(x + 1)$

18. $y - 5 = 3(x - 4)$

19. $y + 11 = -3(x - 9)$

20. $y + 6 = \frac{1}{2}(x - 12)$

21. $y - \frac{2}{3} = 4(x + \frac{5}{12})$

Classified Ads **In Exercises 22 and 23, use the following information.**

It costs $1.50 per day to place a one-line ad in the classifieds plus a flat service fee. One day costs $3.50 and four days costs $8.00.

22. Write a linear equation in point-slope form that gives the cost in dollars, y, in terms of the number of days the ad appears, x.

23. Find the cost of a six-day ad.

Travel **In Exercises 24 and 25, use the following information.**

You are flying from Houston to Chicago. You leave Houston at 7:30 A.M. At 8:35 A.M. you fly over Little Rock, a distance of 455 miles.

24. Write a linear equation that gives the distance in miles, y, in terms of time, x. Let x represent the number of minutes since 7:30 A.M.

25. Approximately what time will you arrive in Chicago if it is 950 miles from Houston?

LESSON 5.2

Reteaching with Practice

For use with pages 276–284

GOAL **Use point-slope form to write the equation of a line.**

VOCABULARY

The **point-slope form** of the equation of the line through (x_1, y_1) with slope m is $y - y_1 = m(x - x_1)$.

Using the Point-Slope Form

EXAMPLE 1

Use the point-slope form of a line to write an equation of the line that passes through the point $(3, -1)$ and has a slope of -1.

SOLUTION

Use the slope of -1 and the point $(3, -1)$ as (x_1, y_1) in the point-slope form.

$y - y_1 = m(x - x_1)$	Write point-slope form.
$y - (-1) = -1(x - 3)$	Substitute –1 for m, 3 for x_1, and –1 for y_1.
$y + 1 = -1(x - 3)$	Simplify.

Using the distributive property, you can write an equation in point-slope form in slope-intercept form.

$y + 1 = -x + 3$	Use distributive property.
$y = -x + 2$	Subtract 1 from each side.

Exercises for Example 1

Write in slope-intercept form the equation of the line that passes through the given point and has the given slope.

1. $(4, 5), m = 2$ **2.** $(-1, 6), m = -3$ **3.** $(-2, 8), m = -4$

---

26   **Algebra 1**
Chapter 5 Resource Book

Reteaching with Practice

For use with pages 276–284

EXAMPLE 2 *Writing an Equation of a Parallel Line*

Write in slope-intercept form the equation of the line that is parallel to the line $y = 3x - 5$ and passes through the point $(-5, -2)$.

SOLUTION

The slope of the original line is $m = 3$. So the slope of the parallel line is also $m = 3$. The line passes through the point $(x_1, y_1) = (-5, -2)$.

$$y - y_1 = m(x - x_1)$$ Write point-slope form.

$$y - (-2) = 3(x - (-5))$$ Substitute -2 for y_1, 3 for m, and -5 for x_1.

$$y + 2 = 3(x + 5)$$ Simplify.

$$y + 2 = 3x + 15$$ Use the distributive property.

$$y = 3x + 13$$ Subtract 2 from each side.

Exercises for Example 2

4. Write in slope-intercept form the equation of the line that is parallel to the line $y = -4x + 1$ and passes through the point $(2, -1)$.

5. Write in slope-intercept form the equation of the line that is parallel to the line $y = -x - 7$ and passes through the point $(-4, -4)$.

Lesson 5.2

NAME _____ DATE _____

Quick Catch-Up for Absent Students

For use with pages 276–284

The items checked below were covered in class on (date missed) _____

Developing Concepts Activity: Point-Slope Form

_____ **Goal:** Develop the point-slope form of the equation of a line.

Lesson 5.2: Point-Slope Form

_____ **Goal:** Use point-slope form to write the equation of a line.

Material Covered:

_____ Student Help: Study Tip

_____ Example 1: Point-Slope Form from a Graph

_____ Student Help: Study Tip

_____ Example 2: Write an Equation in Point-Slope Form

_____ Student Help: Study Tip

_____ Example 3: Use Point-Slope Form

_____ Student Help: More Examples

_____ Example 4: Write an Equation of a Parallel Line

Vocabulary:

point-slope form, p. 278

_____ Other (specify) _____

_____

Homework and Additional Learning Support

_____ Textbook (specify) _pp. 281–284_____

_____ Internet: Extra Examples at www.mcdougallittell.com

_____ *Reteaching with Practice* worksheet (specify exercises)_____

_____ *Personal Student Tutor* for Lesson 5.2

NAME _____ DATE _____

Interdisciplinary Application

For use with pages 276–284

Advertising

BUSINESS Marketing is a major factor in the success or failure of a company's product. While many people associate marketing with advertising, most fail to realize it actually consists of four key areas: the product, its price, the promotional campaign, and the distribution area. Advertising is only one facet of promotion. When advertising a product, a company will usually isolate a target audience or customer. The target is a specific group of people that would like to buy and can afford the product.

There are many products and services that are specifically aimed at the youth of America. Young adults have a lot of discretionary income. Discretionary income is money leftover after paying all necessary expenses. A problem arises when trying to market in a different culture or country.

For instance, Mexico has a large percentage of its population less than 18 years of age, and this age group is expanding. Companies that are successfully selling products to teenagers in the U.S. may expand their product lines into Mexico.

In Exercises 1–3, use the following information.

A company selling blue jeans in Mexico had sales of 45 million pesos when the population of teenagers was 3 million people. The marketing department predicts that sales of blue jeans will increase by 10 million pesos for every million teenagers.

1. Write an equation of the line in point-slope form, letting x represent population and y represent sales in millions of pesos. Use the point (3, 45) and a slope of 10.

2. Graph the line in Exercise 1.

3. Predict sales when the teenage population reaches 7 million.

In Exercises 4–6, use the following information.

A music company sells compact discs and cassettes and has sales of 15 million pesos when the teenage population is 3 million. The marketing department predicts that sales of compact discs and cassettes will increase by 7.3 million pesos per million teenagers.

4. Write an equation of the line in point-slope form, letting x represent population and y represent sales in millions of pesos. Use the point (3, 15) and a slope of 7.3.

5. Graph the line found in Exercise 4.

6. Predict sales when the teenage population reaches 6 million.

NAME _____ DATE _____

Challenge: Skills and Applications

For use with pages 276–284

In Exercises 1–8, write an equation in point-slope form of the line that passes through the given points.

1. $\left(\frac{1}{2}, -4\right), (2, 11)$

2. $(8, 3), \left(-\frac{1}{3}, -2\right)$

3. $(-0.5, 0.9), (-3.3, -0.5)$

4. $(3.2, -1.4), (2.4, 1.8)$

5. $\left(\frac{3}{2}, -\frac{1}{3}\right), \left(-\frac{2}{3}, 4\right)$

6. $(5, -4), \left(\frac{1}{2}, -\frac{1}{4}\right)$

7. $(p, q), (-p, 2q)$

8. $(2p, -q), (p, p - q)$

In Exercises 9–12, use the following information.

A line passes through the point $(6, 3)$ and has slope $-\frac{5}{2}$.

9. Write an equation of the line in point-slope form.

10. For the given point $(6, 3)$, the x-coordinate is twice the y-coordinate. Find a point on the line for which the y-coordinate is twice the x-coordinate. Explain your method.

11. Find a point on the line for which the two coordinates are opposites.

12. Find a point on the line for which the sum of the two coordinates is 15.

In Exercises 13–15, use the following information.

Hiking up a mountain, Zahara looked at a topographical map and saw that at 1:00 P.M. her elevation was 5620 feet above sea level. By 2:30 P.M. she had reached an elevation of 6040 feet above sea level.

13. Write an equation in point-slope form that gives Zahara's elevation y, at a time x hours after noon.

14. If Zahara continues at the same rate, what elevation can she expect to reach by 5:00 P.M.?

15. Use the model from Exercise 13 to find Zahara's elevation at 10:00 A.M.

Lesson 5.2

NAME ——————————————————————— DATE ————

Quiz 1

For use after Lessons 5.1–5.2

1. Write an equation of the line whose slope is $m = -3$ and whose
y-intercept is $b = 7$. *(Lesson 5.1)*

2. Write an equation using slope-intercept form of the line shown in the
graph. *(Lesson 5.1)*

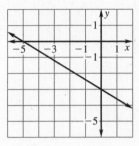

3. Write an equation of the line that passes through $(-3, -9)$ and has a
slope of $m = 4$. Write the equation in slope-intercept form.
(Lesson 5.2)

4. Write an equation in point-slope form of the line shown in the graph.
(Lesson 5.2)

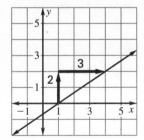

5. Write in slope-intercept form the equation of the line that is parallel to
$y = 2x - 5$ and passes through $(-4, 2)$. *(Lesson 5.2)*

6. Write an equation in slope-intercept form of the line that passes
through $(4, 6)$ and has slope $m = \dfrac{1}{4}$. *(Lesson 5.2)*

7. Write an equation of a line that is parallel to $y = -5x + 2$ and passes
through $(2, 4)$. *(Lesson 5.2)*

Answers

1. _____

2. _____

3. _____

4. _____

5. _____

6. _____

7. _____

Lesson 5.2

TEACHER'S NAME _____ CLASS _____ ROOM _____ DATE _____

Lesson Plan

1-day lesson (See *Pacing the Chapter,* **TE page 266A)**   **For use with pages 285–290**

GOAL **Write an equation of a line given two points on the line.**

State/Local Objectives _____

_____

✓ **Check the items you wish to use for this lesson.**

STARTING OPTIONS
_____ Homework Check (5.2): TE page 281; Answer Transparencies
_____ Homework Quiz (5.2): TE page 283, CRB page 34, or Transparencies
_____ Warm-Up: CRB page 34 or Transparencies

TEACHING OPTIONS
_____ Lesson Opener: CRB page 35 or Transparencies
_____ Examples 1–3: SE pages 285–287
_____ Extra Examples: TE pages 286–287 or Transparencies; Internet Help at *www.mcdougallittell.com*
_____ Checkpoint Exercises: SE pages 285–287
_____ Concept Check: TE page 287
_____ Guided Practice Exercises: SE page 288

APPLY/HOMEWORK
Homework Assignment
_____ Transitional: pp. 288–290, Exs. 9–11, 18–20, 24–26, 33–36, 41, 42, 43–57 odd
_____ Average: pp. 288–290, Exs. 12–14, 21–23, 27–29, 36, 40–42, 44–58 even
_____ Advanced: pp. 288–290, Exs. 15–17, 30–32, 37–42*, 46–49, 53–58; EC: TE p. 266D 1–4, CRB p. 42

Reteaching the Lesson
_____ Practice Masters: CRB pages 36–37 (Level A, Level B)
_____ Reteaching with Practice: CRB pages 38–39 or Practice Workbook with Examples;
       Resources in Spanish
_____ Personal Student Tutor: CD-ROM

Extending the Lesson
_____ Interdisciplinary/Real-Life Applications: CRB page 41
_____ Challenge: CRB page 42

ASSESSMENT OPTIONS
_____ Daily Quiz (5.3): TE page 290, CRB page 45, or Transparencies
_____ Standardized Test Practice: SE page 290; STP Workbook; Transparencies

Notes _____

_____

_____

TEACHER'S NAME _____ CLASS _____ ROOM _____ DATE _____

Lesson Plan for Block Scheduling

Half-block lesson (See *Pacing the Chapter,* TE page 266A)   For use with pages 285–290

GOAL **Write an equation of a line given two points on the line.**

State/Local Objectives _____

_____

_____

CHAPTER PACING GUIDE	
Day	Lesson
1	5.1 (all); 5.2 (begin)
2	5.2 (end); **5.3 (all)**
3	5.4 (all); 5.5 (all)
4	5.6 (all)
5	Ch. 5 Review and Assess

✓ **Check the items you wish to use for this lesson.**

STARTING OPTIONS
_____ Homework Check (5.2): TE page 281; Answer Transparencies
_____ Homework Quiz (5.2): TE page 283,
     CRB page 34, or Transparencies
_____ Warm-Up: CRB page 34 or Transparencies

TEACHING OPTIONS
_____ Lesson Opener: CRB page 35 or Transparencies
_____ Examples 1–3: SE pages 285–287
_____ Extra Examples: TE pages 286–287 or Transparencies; Internet Help at *www.mcdougallittell.com*
_____ Checkpoint Exercises: SE pages 285–287
_____ Concept Check: TE page 287
_____ Guided Practice Exercises: SE page 288

APPLY/HOMEWORK
Homework Assignment (See also the assignment for Lesson 5.2.)
_____ Block Schedule: pp. 288–290: Ex. 12–14, 21–23, 27–29, 33–36, 40, 41, 50–58 even

Reteaching the Lesson
_____ Practice Masters: CRB pages 36–37 (Level A, Level B)
_____ Reteaching with Practice: CRB pages 38–39 or Practice Workbook with Examples;
     Resources in Spanish
_____ Personal Student Tutor: CD-ROM

Extending the Lesson
_____ Interdisciplinary/Real-Life Applications: CRB page 41
_____ Challenge: CRB page 42

ASSESSMENT OPTIONS
_____ Daily Quiz (5.3): TE page 290, CRB page 45, or Transparencies
_____ Standardized Test Practice: SE page 290; STP Workbook; Transparencies

Notes _____

_____

_____

NAME _____ DATE _____

Available as
a transparency

WARM-UP EXERCISES

For use before Lesson 5.3, pages 285–290

Write the equation in slope-intercept form of the line that passes through the point and has the given slope.

1. $(-2, 5), m = \dfrac{1}{2}$ **2.** $(6, -3), m = -2$

Write an equation of the line that is parallel to the given line and passes through the given point.

3. $y = -3x + 2, \ (2, 1)$ **4.** $y = \dfrac{1}{4}x + 1, \ (-2, 0)$

··

DAILY HOMEWORK QUIZ

For use after Lesson 5.2, pages 276–284

1. Write an equation of the line in point-slope form.

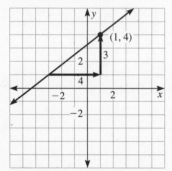

2. Write in point-slope form the equation of the line that passes through the point $(-1, 5)$ and has a slope of -3.

3. Rewrite the equation in point-slope form to an equation in slope-intercept form.

$y - 2 = -4(x + 1)$

4. Write in slope-intercept form an equation of the line that is parallel to the given line and passes through the given point.

$y = \frac{1}{4}x + 3, (-2, -1)$

5. The equation $y = 0.5x + 5.85$ gives your hourly wage y, in dollars per hour, after working at your job for x years. Use the equation to determine your hourly wage after working there for 3 years.

NAME _____ DATE _____

Application Lesson Opener

For use with pages 285–290

**You have a job that pays by the hour. On Monday, you
worked 5 hours and earned $30. On Tuesday, you worked
3 hours and earned $18.**

1. Let x represent the number of hours worked and let y
   represent the amount earned. Write two ordered pairs.

2. Plot the two points given by the ordered pairs. Draw a line
   through the points.

3. Find the slope of the line.

4. Find the y-intercept of the line.

**You use a telephone service that charges $10 a month
plus an hourly fee. One month you paid $4 for 2 hours,
for a total monthly bill of $14. Another month, you paid
$10 for 5 hours, for a total monthly bill of $20.**

5. Let x represent the number of hours and let y represent the
   total monthly bill. Write two ordered pairs.

6. Plot the two points given by the ordered pairs. Draw a line
   through the points.

7. Find the slope of the line.

8. Find the y-intercept of the line.

9. In the two situations above, do you have enough information
   to write equations of the lines? Explain your answer.

Write an equation in slope-intercept form of the line shown in the graph.

1.
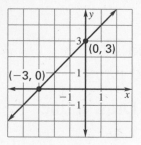

2.
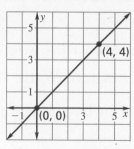

3.
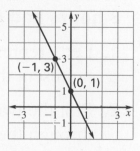

4.
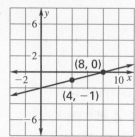

5.
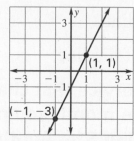

6.
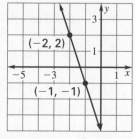

Write an equation in slope-intercept form of the line that passes through the points.

7. $(0, 0), (3, -6)$

8. $(0, 4), (-1, 3)$

9. $(-5, 9), (-2, 0)$

10. $(0, 2), (-2, 0)$

11. $(5, 0), (-10, -5)$

12. $(1, 1), (3, 3)$

13. $(1, -7), (3, -15)$

14. $(-6, -2), (-10, -14)$

15. $(2, 3), (6, 11)$

16. *Learning a Language*   By the end of your 5th French lesson you have learned 20 vocabulary words. Write an equation that gives the number of vocabulary words you know, *y*, in terms of the number of lessons you have had, *x*. Assume you learned the same number of words each lesson.

17. *United Nations*   In 1945, when the United Nations was formed, there were 51 member nations. In 1987, there were 159 member nations. Write an equation that gives the number of nations in the UN, *y*, in terms of the year, *t*. Let *t* = 0 correspond to 1945 and assume that membership followed a linear pattern.

18. *Diving*   Leslie dives off a block at the edge of the pool. She enters the water 8 ft from the side of the pool. Leslie is 1 ft under water when she is 11 ft from the side of the pool. Write an equation that gives Leslie's depth, *y*, in terms of her distance, *x*, from the side.

19. *Nature Hike*   Use the diagram at the right to write the equation of the line from point *A* to point *B*. What is the slope of this line?

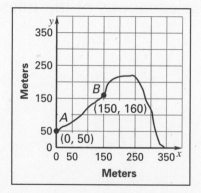

Lesson 5.3

NAME _____ DATE _____

Practice B

For use with pages 285–290

Write an equation in slope-intercept form of the line shown in the graph.

1.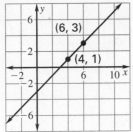

2.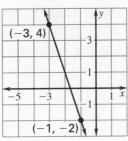

3.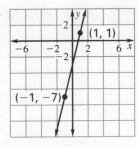

4.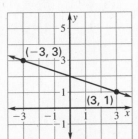

5.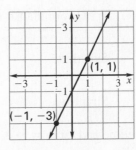

6.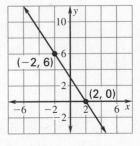

Write an equation in slope-intercept form of the line that passes through the points.

7. $(0, 8), (-1, 3)$

8. $(-7, 9), (-5, -3)$

9. $(3, 2), (7, 5)$

10. $(4, 2), (3, 5)$

11. $(-5, -6), (2, 8)$

12. $(-5, 6), (-6, 1)$

13. $\left(\frac{1}{2}, -1\right), \left(3, \frac{3}{2}\right)$

14. $(6.22, -3.75), (-1.78, 0.25)$

15. $\left(\frac{1}{8}, \frac{7}{8}\right), \left(\frac{3}{4}, -\frac{5}{4}\right)$

16. Write equations of the lines passing through the two parallel sides. How do you know mathematically that these two sides are parallel?

17. *Driving* You drove to your cousin's house, which is 460 miles away. After two hours, you had gone 100 miles. After 8 hours, you had reached your destination. Write an equation that gives the number of miles you had driven, *y*, in terms of the number of hours you had driven, *t*.

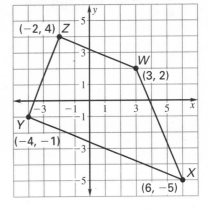

Algebra 1
Chapter 5 Resource Book

37

Lesson 5.3

NAME _____ DATE _____

Reteaching with Practice

For use with pages 285–290

GOAL **Write an equation of a line given two points on the line.**

EXAMPLE 1 *Writing an Equation Given Two Points*

Write an equation of the line that passes through the points $(1, 5)$ and $(2, 3)$.

SOLUTION

Find the slope of the line. Let $(x_1, y_1) = (1, 5)$ and $(x_2, y_2) = (2, 3)$.

$$m = \frac{y_2 - y_1}{x_2 - x_1} \qquad \text{Write formula for slope.}$$

$$= \frac{3 - 5}{2 - 1} \qquad \text{Substitute.}$$

$$= \frac{-2}{1} = -2 \qquad \text{Simplify.}$$

Write the equation of the line and let $m = -2$, $x_1 = 1$, and $y_1 = 5$ and solve for b.

$$y - y_1 = m(x - x_1) \qquad \text{Write point-slope form.}$$

$$y - 5 = -2(x - 1) \qquad \text{Substitute } -2 \text{ for } m, 1 \text{ for } x_1, \text{ and } 5 \text{ for } y_1.$$

$$y = -2x + 7 \qquad \text{Distribute and simplify.}$$

Exercises for Example 1

Write an equation in slope-intercept form of the line that passes through the points.

1. $(4, 9)$ and $(1, 6)$ **2.** $(0, 7)$ and $(1, -1)$ **3.** $(-2, -3)$ and $(0, 3)$

NAME _____ DATE _____

Reteaching with Practice

For use with pages 285–290

EXAMPLE 2 *Decide Which Form to Use*

Write the equation of the line in slope-intercept form.

a.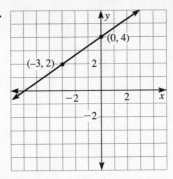

b.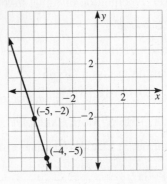

SOLUTION

a. Find the slope.

$$m = \frac{y_2 - y_1}{x_2 - x_1} = \frac{4 - 2}{0 - (-3)} = \frac{2}{3}$$

The y-intercept is $b = 4$.

$$y = mx + b$$

$$y = \frac{2}{3}x + 4$$

b. Find the slope.

$$m = \frac{y_2 - y_1}{x_2 - x_1} = \frac{-5 - (-2)}{-4 - (-5)} = \frac{-3}{1} = -3$$

Since you do not know the y-intercept, use the point-slope form.

$$y - y_1 = m(x - x_1)$$
$$y - (-2) = -3(x - (-5))$$
$$y - 2 = -3x - 15$$
$$y = -3x - 17$$

Exercises for Example 2

Write the equation of the line in slope-intercept form.

4.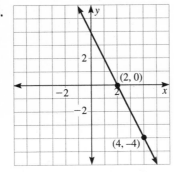

5.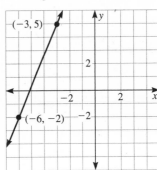

Algebra 1
Chapter 5 Resource Book

39

Lesson 5.3

NAME _____ DATE _____

Quick Catch-Up for Absent Students

For use with pages 285–290

The items checked below were covered in class on (date missed) _____

Lesson 5.3: Writing Linear Equations Given Two Points

_____ **Goal:** Write an equation of a line given two points on the line.

Material Covered:

_____ Student Help: Study Tip

_____ Example 1: Use a Graph

_____ Example 2: Write an Equation of a Line Given Two Points

_____ Student Help: More Examples

_____ Example 3: Decide Which Form to Use

_____ Other (specify) _____

_____

Homework and Additional Learning Support

_____ Textbook (specify) _pp. 288–290_ _____

_____ Internet: Extra Examples at www.mcdougallittell.com

_____ *Reteaching with Practice* worksheet (specify exercises)_____

_____ *Personal Student Tutor* for Lesson 5.3

Lesson 5.3

NAME _____ DATE _____

Interdisciplinary Application

For use with pages 285–290

Bald Eagles

BIOLOGY The bald eagle, our national symbol, is making a comeback from the brink of extinction. Although it has been illegal to hunt bald eagles since 1940, when the Bald Eagle Protection Act made it illegal to kill, harm, harass, or possess bald eagles, the eagle was listed as endangered in 1978 in most of the lower 48 states.

The eagle population was decimated after World War II when the pesticide DDT went into widespread use. The pesticide caused the birds to lay thin-shelled eggs that broke during incubation. DDT was banned in the U.S. on December 31, 1972. Since then, the eagle has steadily increased in numbers, and was down-listed from endangered to threatened in August 1995. In July 1999, the Fish and Wildlife Service proposed removing it from the threatened list under the Endangered Species Act.

The Raptor Research and Technical Assistance Center coordinates the bald eagle survey throughout the lower 48 states. The survey reported 15,896 eagles in 1994. In 1995, 16,289 eagles were counted.

1. Write a linear equation to model the eagle population. Let x represent the number of years since 1990 and y the eagle population.

2. Graph the equation from Exercise 1.

3. Use the equation from Exercise 1 to estimate the eagle population in the year 2000.

4. If the number of eagles counted in 1999 is 18,362, write a linear equation using year 1999 and year 1995 as your points.

5. Graph the equation from Exercise 5.

6. Use the equation from Exercise 5 to estimate the eagle population in the year 2000.

NAME _____ DATE _____

Challenge: Skills and Applications

For use with pages 285–290

In Exercises 1–3, write an equation in slope-intercept form of the line.

   1. through $\left(2\frac{1}{4}, -5\right)$ and $\left(-1\frac{1}{2}, 3\frac{1}{3}\right)$

   2. through $\left(-\frac{1}{6}, \frac{2}{3}\right)$ and parallel to $4x - 2y = 9$

   3. through $(3, -2)$ and parallel to $8 - 3x = 9y$

In Exercises 4–7, use the following information.

Rectangle $ABCD$ has vertices at $A(4, 7)$, $B(3, 1)$, and $C(-3, 2)$.

   4. Find the equation of the line that contains $\overline{AB}$.

   5. Find the equation of the line that contains $\overline{BC}$.

   6. Find the equation of the line that contains $\overline{CD}$.

   7. Find the equation of the line that contains $\overline{AD}$.

In Exercises 8–11, use the following information.

Suppose that a certain type of pea plant requires 14 days to reach a height of 6 inches and 30 days to reach a height of 16 inches.

   8. Write a linear equation that models the height of the plant after x days.

   9. About how many days would it take a plant of this type to reach a height of 12 inches?

   10. What should the height of the plant be after 20 days?

   11. According to the model, what should the height of the plant be after zero days? Why do you think this value is negative?

TEACHER'S NAME _____ CLASS _____ ROOM _____ DATE _____

Lesson Plan

1-day lesson (See *Pacing the Chapter,* TE page 266A)   **For use with pages 291–297**

GOAL **Write an equation of a line in standard form.**

State/Local Objectives _____

_____

✓ Check the items you wish to use for this lesson.

STARTING OPTIONS

____ Homework Check (5.3): TE page 288; Answer Transparencies

____ Homework Quiz (5.3): TE page 290, CRB page 45, or Transparencies

____ Warm-Up: CRB page 45 or Transparencies

TEACHING OPTIONS

____ Lesson Opener: CRB page 46 or Transparencies

____ Examples 1–4: SE pages 291–293

____ Extra Examples: TE pages 292–293 or Transparencies

____ Checkpoint Exercises: SE pages 291–293

____ Concept Check: TE page 293

____ Guided Practice Exercises: SE page 294

APPLY/HOMEWORK

Homework Assignment

____ Transitional: pp. 294–297, Exs. 15–17, 21–23, 30–32, 39, 40, 54–56, 61, 62, 63–79 odd, Quiz 2

____ Average: pp. 294–297, Exs. 18–20, 24–26, 33–35, 41, 42, 53, 61, 62, 64–78 even, Quiz 2

____ Advanced: pp. 294–297, Exs. 27–29, 36–38, 43–52, 57–62*, 66–68, 76–79, Quiz 2; EC: CRB p. 53

Reteaching the Lesson

____ Practice Masters: CRB pages 47–48 (Level A, Level B)

____ Reteaching with Practice: CRB pages 49–50 or Practice Workbook with Examples;
Resources in Spanish

____ Personal Student Tutor: CD-ROM

Extending the Lesson

____ Interdisciplinary/Real-Life Applications: CRB page 52

____ Challenge: CRB page 53

ASSESSMENT OPTIONS

____ Daily Quiz (5.4): TE page 297, CRB page 57, or Transparencies

____ Standardized Test Practice: SE page 296; STP Workbook; Transparencies

____ Quiz 5.3–5.4: SE page 297; CRB page 54; Resources in Spanish

Notes _____

_____

_____

Lesson 5.4

TEACHER'S NAME _____ CLASS _____ ROOM _____ DATE _____

Lesson Plan for Block Scheduling

Half-block lesson (See *Pacing the Chapter,* TE page 266A)   For use with pages 291–297

GOAL   Write an equation of a line in standard form.

State/Local Objectives _____

_____

_____

✓ **Check the items you wish to use for this lesson.**

STARTING OPTIONS

____ Homework Check (5.3): TE page 288; Answer Transparencies
____ Homework Quiz (5.3): TE page 290,
CRB page 45, or Transparencies
____ Warm-Up: CRB page 45 or Transparencies

CHAPTER PACING GUIDE	
Day	**Lesson**
1	5.1 (all); 5.2 (begin)
2	5.2 (end); 5.3 (all)
3	**5.4 (all)**; 5.5 (all)
4	5.6 (all)
5	Ch. 5 Review and Assess

TEACHING OPTIONS

____ Lesson Opener: CRB page 46 or Transparencies
____ Examples 1–4: SE pages 291–293
____ Extra Examples: TE pages 292–293 or Transparencies
____ Checkpoint Exercises: SE pages 291–293
____ Concept Check: TE page 293
____ Guided Practice Exercises: SE page 294

APPLY/HOMEWORK

Homework Assignment (See also the assignment for Lesson 5.5.)

____ Block Schedule: pp. 294–297: Exs. 18–20, 24–26, 33–35, 41, 42, 53, 61, 62, 64–78 even, Quiz 2

Reteaching the Lesson

____ Practice Masters: CRB pages 47–48 (Level A, Level B)
____ Reteaching with Practice: CRB pages 49–50 or Practice Workbook with Examples;
Resources in Spanish
____ Personal Student Tutor: CD-ROM

Extending the Lesson

____ Interdisciplinary/Real-Life Applications: CRB page 52
____ Challenge: CRB page 53

ASSESSMENT OPTIONS

____ Daily Quiz (5.4): TE page 297, CRB page 57, or Transparencies
____ Standardized Test Practice: SE page 296; STP Workbook; Transparencies
____ Quiz 5.3–5.4: SE page 297; CRB page 54; Resources in Spanish

Notes _____

_____

_____

NAME _____ DATE _____

WARM-UP EXERCISES

For use before Lesson 5.4, pages 291–297

Write an equation for each line in point-slope form.

1. the line through $(-1, 3)$ with slope -2

2. the line through $(3, -5)$ and $(2, -4)$

**Convert each equation in point-slope form to
slope-intercept form.**

3. $y + 1 = -2(x - 5)$ **4.** $y - 3 = \dfrac{1}{2}(x + 6)$

DAILY HOMEWORK QUIZ

For use after Lesson 5.3, pages 285–290

1. Write in point-slope form the equation of the line that
passes through the points $(4, 3)$ and $(1, 2)$.

2. Write the equation of the
line in slope-intercept form.

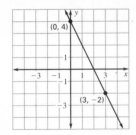

3. Write in slope-intercept
form the equation of the
line that passes through the
points $(4, 5)$ and $(1, -1)$.

4. The graph below models the distance a person walks.
Write in slope-intercept form the equation of the line.

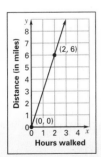

Algebra 1
Chapter 5 Resource Book

45

SET UP: Work with a group.

YOU WILL NEED: • **16 index cards**

1. Write each equation on a separate index card.

$y = 2x + 1$ $-x + 5y = 10$

$y = 4x - 1$ $x - y = 6$

$y = \dfrac{1}{5}x + 2$ $x + 2y = -8$

$y = -x + 3$ $-2x + y = 1$

$y = -\dfrac{1}{2}x - 4$ $x + y = 3$

$y = 5x + 1$ $4x - y = 1$

$y = x - 6$ $5x - y = -1$

$y = \dfrac{2}{3}x + 3$ $-2x + 3y = 9$

2. Use your cards to play "Equation Concentration." Shuffle the cards and place them face down on a desk or table. Choose a member of your group to go first. The first player turns over two cards. If the player thinks that the equations on the cards represent equivalent equations, he or she picks up the two cards. If not, he or she turns the cards over and returns them to the bottom of the deck. The player to the right goes next. If two cards are picked up and any other player thinks the equations on the two cards are not equivalent, he or she can call for a "challenge." To win a "challenge," a player must prove that the equations are not equivalent. The player who loses a "challenge" loses his or her next turn. A player's turn ends when he or she loses a challenge or when two cards are turned over that do not match. The game is over when all the cards have been matched. The player with the most cards at the end of the game "wins."

Practice A
For use with pages 291–297

Write the equation in standard form with integer coefficients.

1. $x - y - 9 = 0$

2. $-4y + 6x + 7 = 0$

3. $x + 7 = 0$

4. $3y + 2x = 6$

5. $y = -11x - 4$

6. $y - 1 = 0$

7. $3 + 4x - y = 0$

8. $x - 8y + 2 = 0$

9. $x = y$

10. $y = 5x - \frac{1}{2}$

11. $y = \frac{1}{4}x + 3$

12. $y = -\frac{2}{3}x - 1$

Write the standard form of the equation of the line that passes through the given point and has the given slope.

13. $(0, 4), m = 1$

14. $(2, 5), m = -3$

15. $(-1, 3), m = 8$

16. $(6, -7), m = 4$

17. $(5, 6), m = -2$

18. $(-4, -9), m = -2$

Write the standard form of the equation of the line that passes through the given points.

19. $(2, -5), (8, 1)$

20. $(-1, -2), (0, 3)$

21. $(1, -6), (-5, 6)$

22. $(3, 19), (-2, -11)$

23. $(-4, 3), (-1, -6)$

24. $(2, 18), (-2, 2)$

25. *Publicity* You are running for class president. You have $30 to spend on publicity. It costs $2 to make a campaign button and $1 to make a poster. Write an equation that represents the different numbers of buttons, *x,* and posters, *y,* you could make.

26. Sketch the line representing the possible combinations of buttons and posters in Exercise 25. Then complete the table and label the points from the table on the graph.

Number of buttons	0	5	8	10	15
Number of posters					

27. *Canning Jelly* Your grandmother made 240 oz of jelly. You have two types of jars. The first holds 10 oz and the second holds 12 oz. Write an equation that represents the different numbers of 10-oz jars, *x,* and 12-oz jars, *y,* that will hold all of the jelly.

12 oz   **10 oz**

28. Sketch the line representing the possible jar combinations in Exercise 27. Then complete the table and label the points from the table on the graph.

10-oz jars	0	6	12	18	24
12-oz jars					

NAME _____ DATE _____

Practice B

For use with pages 291–297

Write the equation in standard form with integer coefficients.

1. $2x - y - 8 = 0$
2. $0.3x - 0.4y = 7.5$
3. $y = 3x + 2$
4. $y = 5 - 3x$
5. $0.6x = 2.1y + 1.8$
6. $2x = 3y + 5$
7. $x - 4 = 0$
8. $3y = 12$
9. $2x - 9 = \frac{3}{5}y$
10. $\frac{1}{4}x - 2y = -3$
11. $y = \frac{1}{2}x + 4$
12. $y = \frac{2}{3}x - \frac{5}{3}$

Write the standard form of the equation of the line that passes through the given point and has the given slope.

13. $(4, 3), m = 2$
14. $(1, 5), m = -4$
15. $(0, 6), m = 3$
16. $(-2, 4), m = -6$
17. $(6, -8), m = \frac{1}{3}$
18. $(-2, 4), m = -\frac{1}{2}$

Write the standard form of the equation of the line that passes through the given points.

19. $(5, 8), (3, 2)$
20. $(-2, 5), (3, -10)$
21. $(-7, 3), (1, 2)$
22. $(-4, -5), (-2, 5)$
23. $(8, 1), (4, -1)$
24. $(-6, 6), (3, 3)$

Write the standard form of the equation of the horizontal and vertical lines that pass through the given point.

25. $(3, -4)$
26. $(5, 1)$
27. $(-3, -2)$
28. $(0, -4)$

Party Food **In Exercises 29–32, use the following information.**

You are in charge of buying the hamburger and boned chicken for a party. You have $60 to spend. The hamburger costs $2 per pound and boned chicken is $3 per pound.

29. Write an equation that represents the different amounts of hamburger, x, and chicken, y, that you can buy.

30. Rewrite the equation in Exercise 29 in slope-intercept form.

31. Sketch the graph of the linear equation in Exercise 29.

32. Complete the table and label the points from the table on the graph.

Hamburger (lb), x	0	6	12	18	30
Chicken (lb), y					

Lawn Seed **In Exercises 33–36, use the following information.**

You are buying $48 worth of lawn seed that consists of two types of seed. One type is a quick-growing rye grass that costs $4 per pound, and the other type is a higher-quality seed that costs $6 per pound.

33. Write an equation that represents the different amounts of $4 seed, x, and $6 seed, y, that you can buy.

34. Rewrite the equation in Exercise 33 in slope-intercept form.

35. Sketch the graph of the linear equation in Exercise 33.

36. Complete the table and label the points from the table on the graph.

$4 seed (lb), x	0	3	6	9	12
$6 seed (lb), y					

NAME _____ DATE _____

Reteaching with Practice

For use with pages 291–297

GOAL **Write an equation of a line in standard form.**

> ### VOCABULARY
>
> The **standard form** of the equation of a line, is $Ax + By = C$, where A and B are not both zero.

EXAMPLE 1 *Converting to Standard Form*

Write $y = -\frac{3}{4}x + 5$ in standard form with integer coefficients.

SOLUTION

To write the equation in standard form, isolate the variable terms on the left and the constant term on the right.

$y = -\frac{3}{4}x + 5$	Write original equation.
$4y = 4\left(-\frac{3}{4}x + 5\right)$	Multiply each side by 4.
$4y = -3x + 20$	Use distributive property.
$3x + 4y = 20$	Add $3x$ to each side.

Exercises for Example 1

Write the equation in standard form with integer coefficients.

1. $y = \frac{2}{3}x - 7$　　　　**2.** $y = 8 + 2x$　　　　**3.** $y = 6 - \frac{1}{4}x$

EXAMPLE 2 *Writing a Linear Equation in Standard Form*

Write the standard form of the equation passing through $(3, 7)$ with a slope of 2.

SOLUTION

Write the point-slope form of the equation of the line.

$y - y_1 = m(x - x_1)$	Write point-slope form.
$y - 7 = 2(x - 3)$	Substitute 7 for y_1, 2 for m, and 3 for x_1.
$y - 7 = 2x - 6$	Use distributive property.
$y = 2x + 1$	Add 7 to each side.
$-2x + y = 1$	Subtract $2x$ from each side.

Reteaching with Practice

For use with pages 291–297

Exercises for Example 2

Write the standard form of the equation of the line that passes through the given point and has the given slope.

4. $(1, 4)$, $m = -2$ **5.** $(-3, 1)$, $m = 3$ **6.** $(5, -2)$, $m = -1$

EXAMPLE 3 *Writing an Equation in Standard Form*

A line passes through the points $(0, 2)$ and $(-4, -1)$. Write an equation of the line in standard form. Use integer coefficients.

❶ Find the slope. Use $(x_1, y_1) = (0, 2)$ and $(x_2, y_2) = (-4, -1)$.

$$m = \frac{y_2 - y_1}{x_2 - x_1} = \frac{-1 - 2}{-4 - 0} = \frac{-3}{-4} = \frac{3}{4}$$

❷ Write an equation of the line, using slope-intercept form.

$y = mx + b$	Write slope-intercept form.
$y = \dfrac{3}{4}x + 2$	Substitute $\dfrac{3}{4}$ for m and 2 for b.
$4y = 4\left(\dfrac{3}{4}x + 2\right)$	Multiply each side by 4.
$4y = 3x + 8$	Use distributive property.
$-3x + 4y = 8$	Subtract $3x$ from each side.

Exercises for Example 3

7. Write in standard form an equation of the line that passes through the points $(1, 3)$ and $(0, 8)$. Use integer coefficients.

NAME _____ DATE _____

Quick Catch-Up for Absent Students

For use with pages 291–297

The items checked below were covered in class on (date missed) _____

Lesson 5.4: Standard Form

_____ **Goal:** Write an equation of a line in standard form.

Material Covered:

_____ Student Help: Vocabulary Tip

_____ Example 1: Convert to Standard Form

_____ Student Help: Study Tip

_____ Example 2: Write an Equation in Standard Form

_____ Student Help: Study Tip

_____ Example 3: Write an Equation in Standard Form

_____ Student Help: Study Tip

_____ Example 4: Equations for Horizontal and Vertical Lines

Vocabulary:

standard form, p. 291

_____ Other (specify) _____

_____

Homework and Additional Learning Support

_____ Textbook (specify) __pp. 294–297_____

_____ *Reteaching with Practice* worksheet (specify exercises)_____

_____ *Personal Student Tutor* for Lesson 5.4

Real-Life Application:
When Will I Ever Use This?

For use with pages 291–297

Saving Money

The best way to get in the habit of saving money is to start young and be consistent. Even a small amount of money saved each month will add up over time. Once people start saving, most quickly realize they do not miss the money.

In Exercises 1–3, use the following information.

Mitch wants to save some money to purchase his own car. He gets two part-time jobs during summer vacation. The first, working as a busboy at a local restaurant, pays $8 per hour. The second, a gas station attendant position, pays $10 per hour. He would like to earn $200 per week.

1. Write an equation in standard form that models the different amounts of time he can work each job. Let x be the number of hours worked at the restaurant and y be the number of hours worked at the gas station.

2. Graph the equation.

3. If Mitch saves 60% of his income, how much will he have saved by the end of 12 weeks of summer vacation?

In Exercises 4–5, use the following information.

Halie wants a phone line in her room. Her parents say she can have her own phone line if she pays for the installation, the phone, and the monthly bill. They decide she must have $300 saved to cover all these costs. She earns $3 an hour babysitting and $2 an hour for household chores.

4. Write an equation in standard form that models the number of hours babysitting b and hours of chores c completed to raise the required money.

5. Graph the equation. Put b on the horizontal axis and c on the vertical axis.

Lesson 5.4

NAME _____ DATE _____

Challenge: Skills and Applications

For use with pages 291–297

In Exercises 1–4, write an equation in standard form of the line that passes through the two points.

Example: $\left(\frac{1}{4}, \frac{3}{8}\right), \left(\frac{3}{5}, \frac{1}{2}\right)$

Solution:	$y - \frac{1}{2} = \frac{5}{14}\left(x - \frac{3}{5}\right)$	Find slope and write equation in point-slope form.
	$y - \frac{1}{2} = \frac{5}{14}x - \frac{3}{14}$	Use distributive property.
	$14\left(y - \frac{1}{2}\right) = 14\left(\frac{5}{14}x - \frac{3}{14}\right)$	Multiply each side by least common denominator.
	$14y - 7 = 5x - 3$	Use distributive property.
	$-5x + 14y = 4$	Add $-5x$ and 7 to each side.

1. $\left(\frac{1}{2}, -\frac{2}{3}\right), \left(\frac{3}{4}, \frac{7}{8}\right)$ **2.** $\left(-3, 2\frac{1}{4}\right), \left(3\frac{1}{5}, 10\right)$

3. $(-1, p), (2, -4)$ **4.** $(3, -2), (1, q)$

In Exercises 5–7, use the following information.

Melissa is trying to determine the composition of a 12-milligram chemical solution. She knows the solution contains two chemicals. Each milliliter of one chemical weighs 2 milligrams and each milliliter of the other weighs 3 milligrams.

5. Write an equation that represents the different numbers of milliliters of each chemical that could be in the solution.

6. If there are 1.8 milliliters of the first chemical, how much of the second chemical is there?

7. Is it possible that the mixture contains the same number of milliliters of each chemical? If so, what is that number of milliliters?

In Exercises 8–9, Steven is cutting shelves that are either 4.5 feet or 3.75 feet long. (Ignore the width of the saw blade.)

8. Write an equation that represents the numbers of shelves of each length that Steven can cut if he has 24 feet of wood. How many shelves of each length can he cut with no wood left over?

9. The wood is only sold in lengths of 12 feet or less. Will Steven need to piece together sections of wood to make a shelf? Explain.

NAME _____ DATE _____

Quiz 2

For use after Lessons 5.3–5.4

Write in slope-intercept form the equation of the line that passes through the two points. *(Lesson 5.3)*

1. (3, 4) and (2, 6)

2. (−1, −2) and (0, 3)

3. Write an equation of the line in standard form. *(Lesson 5.4)*

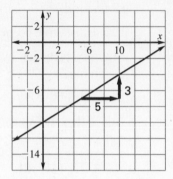

Answers

1. _____

2. _____

3. _____

4. _____

5. _____

4. Write an equation in standard form of the line that passes through (2, 7) and (−6, 3). *(Lesson 5.4)*

5. Write an equation in standard form of the line that passes through (−8, 1) and has a slope of $m = \frac{1}{2}$. *(Lesson 5.4)*

Algebra 1
Chapter 5 Resource Book

TEACHER'S NAME _____ CLASS _____ ROOM _____ DATE _____

Lesson Plan

1-day lesson (See *Pacing the Chapter,* TE page 266A)   **For use with pages 298–304**

GOAL **Write and use a linear equation to solve a real-life problem.**

State/Local Objectives _____

_____

✓ **Check the items you wish to use for this lesson.**

STARTING OPTIONS
____ Homework Check (5.4): TE page 294; Answer Transparencies
____ Homework Quiz (5.4): TE page 297, CRB page 57, or Transparencies
____ Warm-Up: CRB page 57 or Transparencies

TEACHING OPTIONS
____ Lesson Opener: CRB page 58 or Transparencies
____ Examples 1–3: SE pages 298–300
____ Extra Examples: TE pages 299–300 or Transparencies
____ Checkpoint Exercises: SE pages 298–300
____ Concept Check: TE page 300
____ Guided Practice Exercises: SE page 301

APPLY/HOMEWORK
Homework Assignment
____ Transitional: EP p. 40, Exs. 16–20; pp. 301–304, Exs. 7–11, 24–26, 35, 36, 37–57 odd
____ Average: pp. 301–304, Exs. 12–17, 31, 32, 35, 36, 38–58 even
____ Advanced: pp. 301–304, Exs. 18–23, 27–30, 33–36, 40–42, 53–58; EC: TE p. 266D a–d, CRB p. 66

Reteaching the Lesson
____ Practice Masters: CRB pages 59–60 (Level A, Level B)
____ Reteaching with Practice: CRB pages 61–62 or Practice Workbook with Examples;
      Resources in Spanish
____ Personal Student Tutor: CD-ROM

Extending the Lesson
____ Learning Activity: CRB page 64
____ Interdisciplinary/Real-Life Applications: CRB page 65
____ Challenge: CRB page 66

ASSESSMENT OPTIONS
____ Daily Quiz (5.5): TE page 304, CRB page 69, or Transparencies
____ Standardized Test Practice: SE page 304; STP Workbook; Transparencies

Notes _____

_____

_____

Lesson Plan for Block Scheduling

Half-block lesson (See *Pacing the Chapter,* TE page 266A) **For use with pages 298–304**

GOAL **Write and use a linear equation to solve a real-life problem.**

State/Local Objectives _____

_____

_____

✓ **Check the items you wish to use for this lesson.**

STARTING OPTIONS

_____ Homework Check (5.4): TE page 294; Answer Transparencies

_____ Homework Quiz (5.4): TE page 297,
CRB page 57, or Transparencies

_____ Warm-Up: CRB page 57 or Transparencies

TEACHING OPTIONS

_____ Lesson Opener: CRB page 58 or Transparencies

_____ Examples 1–3: SE pages 298–300

_____ Extra Examples: TE pages 299–300 or Transparencies

_____ Checkpoint Exercises: SE pages 298–300

_____ Concept Check: TE page 300

_____ Guided Practice Exercises: SE page 301

APPLY/HOMEWORK

Homework Assignment (See also the assignment for Lesson 5.4.)

_____ Block Schedule: pp. 301–304 Exs. 12–17, 31, 32, 38–58 even

Reteaching the Lesson

_____ Practice Masters: CRB pages 59–60 (Level A, Level B)

_____ Reteaching with Practice: CRB pages 61–62 or Practice Workbook with Examples;
Resources in Spanish

_____ Personal Student Tutor: CD-ROM

Extending the Lesson

_____ Learning Activity: CRB page 64

_____ Interdisciplinary/Real-Life Applications: CRB page 65

_____ Challenge: CRB page 66

ASSESSMENT OPTIONS

_____ Daily Quiz (5.5): TE page 304, CRB page 69, or Transparencies

_____ Standardized Test Practice: SE page 304; STP Workbook; Transparencies

Notes _____

_____

_____

CHAPTER PACING GUIDE	
Day	**Lesson**
1	5.1 (all); 5.2 (begin)
2	5.2 (end); 5.3 (all)
3	5.4 (all); **5.5 (all)**
4	5.6 (all)
5	Ch. 5 Review and Assess

LESSON 5.5

NAME _____ DATE _____

WARM-UP EXERCISES

For use before Lesson 5.5, pages 298–304

Write a linear equation in slope-intercept form for the line through the two points.

1. (2, 10) and (6, 18)

2. (−1, 3) and (0, 5)

3. (−4, −4) and (4, 12)

4. Using the pattern of equations in Exercises 1–3, write the next equation in the series.

··

DAILY HOMEWORK QUIZ

For use after Lesson 5.4, pages 291–297

1. Write in standard form an equation of the line. Use integer coefficients.
   $y = 3x + \frac{1}{4}$

2. Write in standard form an equation of the line that passes through (−3, 4) and has a slope of −2.

3. Write in standard form the equation of the line that passes through the points (−6, 0) and (0, 12).

4. Write in standard form an equation of the line.

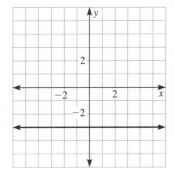

5. You are buying two kinds of trading cards. The first type costs $2 and the second type costs $1. You have only $8 to spend. The equation $2x + y = 8$ models the number of the first type x and the second type y that you can buy. Graph the line modeling the numbers of trading cards you can buy.

NAME _____ DATE _____

Visual Approach Lesson Opener

For use with pages 298–304

Write a linear equation in standard form the line shown.

1.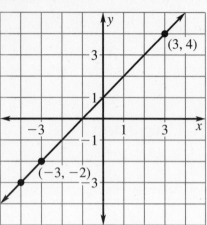

2.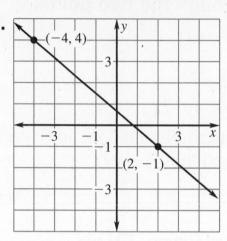

3.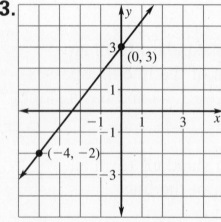

4.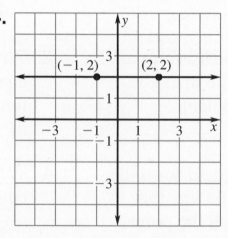

5.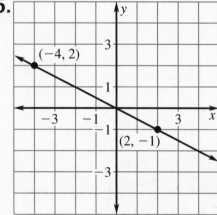

6.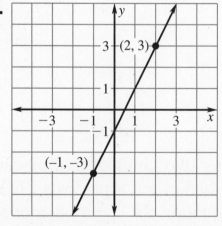

Algebra 1
Chapter 5 Resource Book

NAME _____ DATE _____

Practice A

For use with pages 298–304

Boating   **In Exercises 1–5, use the following information.**

You want to rent a rowboat for a fishing trip. It costs $8 plus $12 per day. The linear model for this situation relates the total cost of renting a rowboat, y, with the number of days rented, x.

1. What number corresponds to the slope in the linear model?

2. What number corresponds to the y-intercept in the linear model?

3. Use the slope and y-intercept to write the linear model.

4. Use the linear model to find the cost of renting a rowboat for 6 days.

5. If you had $50 to spend, for how many days could you rent the rowboat?

Saving   **In Exercises 6–8, use the following information.**

Let $y = 55x + 26$ represent the amount of money (in dollars) in your savings account from 1988 to 1998. Let x represent the number of years since 1988.

6. What is the rate of change in the linear model?

7. Estimate the amount of money in your savings account for 1992.

8. Estimate the amount of money in your savings account for 1998.

Movie Prices   **In Exercises 9–11, use the following information.**

Let $y = 0.25x + 4$ for the cost of going to a movie from 1985 to 1995. Let x represent the number of years since 1985.

9. What is the y-intercept in the linear model?

10. Estimate the cost of going to the movies in 1991.

11. Estimate the cost of going to the movies in 1997.

Gardening   **In Exercises 12–14, use the following information.**

You have $12 to buy tomato and pepper seedlings for your garden. The tomato seedlings cost $4.00 and the pepper seedlings cost $2.00.

12. Write an equation in standard form that represents the different amounts tomato and pepper seedlings that you could buy.

13. Copy the table. Then use the linear equation to complete the table.

Number of tomato seedlings	0	1	2	3
Number of pepper seedlings	?	?	?	?

14. Describe the relationship between the number of tomato seedlings and the number of pepper seedlings shown in the table.

NAME _____ DATE _____

Practice B
For use with pages 298–304

Music **In Exercises 1–5, use the following information.**

You would like to record a song in a music studio. It costs an initial fee of $50 plus $75 per hour. The linear model for this situation relates the total cost of recording a song, y, with the number of hours spent in the studio, x.

1. What number corresponds to the slope in the linear model for the situation?

2. What number corresponds to the y-intercept in the linear model?

3. Use the slope and y-intercept to write the linear model.

4. Use the linear model to find the cost of recording for 3 hours.

5. If you had $425 to spend, for how many hours could you record?

Profit **In Exercises 6–8, use the following information.**

Let $y = 4.2x + 7.1$ represent a company's profit (in millions of dollars) from 1985 to 1995. Let x represent the number of years since 1985.

6. What is the rate of change in the linear model?

7. Estimate the company's profit in 1992.

8. Estimate the company's profit in 1995.

Salary **In Exercises 9–11, use the following information.**

Let $y = 0.3x + 35$ represent a chemist's salary (in thousands of dollars) from 1985 to 1995. Let x represent the number of years since 1985.

9. What is the y-intercept in the linear model?

10. Estimate the chemist's salary in 1991.

11. Estimate the chemist's salary in 1995.

Fruit **In Exercises 12–14, use the following information.**

You have $9.00 to buy fruit for a fresh fruit tray. Grapes cost $3.00 per pound and plums cost $1.50 per pound.

12. Write an equation in standard form that represents the different amounts of pounds of grapes and plums you could buy.

13. Copy the table. Then use the linear equation to complete the table.

Number of pounds of grapes	0	1	2	3
Number of pounds of plums	?	?	?	?

14. Describe the relationship between the number of pounds of grapes and the number of pounds of plums shown in the table.

NAME _____ DATE _____

Reteaching with Practice

For use with pages 298–304

GOAL **Write and use a linear equation to solve a real-life problem.**

VOCABULARY

A **linear model** is a linear function that is used to model a real-life situation.

A **rate of change** compares two quantities that are changing.

EXAMPLE 1 *Write a Linear Model*

In 1995 you had an investment worth $1000. The investment decreased in value by about $50 per year. Write a linear model for the value of your investment y. Let $t = 0$ represent 1995.

SOLUTION

The rate of decrease is $50 per year, so the slope is $m = -50$. The year 1995 is represented by $t = 0$. Therefore, the point (t_1, y_1) is $(0, 1000)$.

$y - y_1 = m(t - t_1)$	Write the point-slope form.
$y - 1000 = -50(t - 0)$	Substitute values.
$y - 1000 = -50t$	Use the distributive property
$y = -50t + 1000$	Add 1000 to each side.

Exercises for Example 1

1. You begin a hiking trail at 8:00 A.M. and hike at a rate of 3 miles per hour. Write a linear model for the number of miles hiked y. Let $t = 0$ represent 8:00 A.M.

2. You make an initial investment of $500 in 1995. It increases in value by about $100 per year. Write a linear model for the value of the investment in year t. Let $t = 0$ represent 1995.

EXAMPLE 2 *Use a Linear Model to Predict*

Use the linear model in Example 1 to predict the value of your investment in 2003.

$y = -50t + 1000$	Write the linear model.
$y = -50(8) + 1000$	Substitute 8 for t.
$y = -400 + 1000$	Simplify.
$y = 600$	Solve for y.

In 2003 your investment will be worth $600.

Reteaching with Practice

For use with pages 298–304

Exercises for Example 2

3. Use the linear model you wrote in Exercise 1 to predict how many miles you will have hiked at 10 A.M. if you continue at the same rate.

4. Use the linear model you wrote in Exercise 2 to predict the value of your investment in 2001 if it increases at the same rate.

EXAMPLE 3 Write and Use a Linear Model

You are buying carrots and peas for dinner. The carrots cost $1.50 per pound and the peas cost $.75. You have $4.50 to spend.

a. Write an equation that models the different amounts (in pounds) of carrots and peas you can buy.

b. Use the model to complete the table that illustrates several different amounts of carrots and peas you can buy.

Carrots (lb), x	0	1	2	3
Peas (lb), y	?	?	?	?

SOLUTION

a. Let the amount of carrots (in pounds) be x and the amount of peas (in pounds) be y.

Verbal Model	Price of carrots	$\cdot$	Weight of carrots	$+$	Price of peas	$\cdot$	Weight of peas	$=$	Total cost

Labels   Price of carrots $= 1.5$ (dollars per pound)

Weight of carrots $= x$ (pounds)

Price of peas $= 0.75$ (dollars per pound)

Weight of peas $= y$ (pounds)

Total cost $= 4.50$ (dollars)

Algebraic Model $1.5x + 0.75y = 4.50$

b. Complete the table by substituting the given values of x into the equation $1.5x + 0.75y = 4.50$ to find y.

Carrots (lb), x	0	1	2	3
Peas (lb), y	6	4	2	0

Exercises for Example 3

5. You are buying jeans and shirts. You have $120. Jeans cost $40 and shirts cost $20. Write an equation that models the different amounts of jeans and shirts you can afford to buy. Use a table to show the different combinations of jeans and shirts you can buy.

NAME _____ DATE _____

Quick Catch-Up for Absent Students

For use with pages 298–304

The items checked below were covered in class on (date missed) _____

Lesson 5.5: Modeling with Linear Equations

____ **Goal:** Write and use a linear equation to solve a real-life problem.

Material Covered:

    ____ Student Help: Study Tip

    ____ Example 1: Write a Linear Model

    ____ Student Help: Study Tip

    ____ Example 2: Use a Linear Model to Predict

    ____ Student Help: Look Back

    ____ Example 3: Write and Use a Linear Model

Vocabulary:

    linear model, p. 298   rate of change, p. 298

____ Other (specify) _____

_____

Homework and Additional Learning Support

    ____ Textbook (specify) <u>pp. 301–304</u> _____

    ____ *Reteaching with Practice* worksheet (specify exercises)_____

    ____ *Personal Student Tutor* for Lesson 5.5

NAME _____ DATE _____

Learning Activity

For use with pages 298–304

GOAL **To make a scatterplot of data and to find a linear model for the data.**

Materials: graph paper, pencil, ruler or meter stick

Exploring Scatter Plots and Linear Models

In this activity, you and your partner will collect data from students in your class. After plotting the data on graph paper, you will find equations of various linear models for the data.

Instructions

1 Measure the heights of ten students in your class and the lengths of their forearms.

2 Graph the data. Use height as the *x*-value and forearm length as the *y*-value.

3 Select two data points and write the equation of the line that passes through them. Graph this line on the scatter plot.

Analyzing the Results

1. Does the line you graphed in Step 3 come reasonably close to the other data points on the scatter plot? If so, use it in Exercise 2. If not, use two other points to create a linear model that better fits the data points.

2. Use your linear model to make predictions about forearm lengths for people who are shorter or taller than those in your data set.

NAME _____ DATE _____

Real-Life Application:
When Will I Ever Use This?

For use with pages 298–304

The Internet

The ideas and technology that led to the development of today's Internet first appeared in the 1960s. However, it was about twenty years until the Internet became widely accessible to people outside universities and scientific and government centers.

The Internet allows computers to directly communicate with each other using many kinds of electronic transports including satellite systems, telephone lines, and optical filters. As more people sign onto the Internet, more hosts, or Internet service providers, will be needed. The Internet Software Consortium published the following survey counting the number of Internet hosts.

Year	Number of Internet Hosts (in Thousands)
1991	376
1992	727
1993	1,313
1994	2,217
1995	4,852
1996	9,472
1997	16,146
1998	29,670
1999	43,230

1. A linear model that approximates the number of Internet hosts based on the 1995–1997 data is $y = 5647x + 4510$, where $x = 0$ represents 1995. Use the model to estimate the number of hosts in 1996.

2. A linear model that approximates the number of Internet hosts based on the 1997–1999 data is $y = 13,542x + 16,140$, where $x = 0$ represents 1997. Use the linear model to estimate the number of hosts in 1998.

3. Which linear model is more accurate?

4. Use the more accurate model to predict the number of Internet hosts in 2003.

Challenge: Skills and Applications

For use with pages 298–304

Sometimes when you create a scatter plot you can write a linear model to approximate the data. There are several ways to do this. In Exercises 1–2, you will write a linear model for the data set by finding the *median-median* line as explained in the steps below. The *median* is the middle number in a data set when the data are in numerical order.

Step 1: Order the data points so that the *x*-values increase from least to greatest.

Step 2: Group the ordered data into three sets, each containing the same number of points. Find the median of the *x*'s and the median of the *y*'s in each set and write them as ordered pairs $(x_1, y_1), (x_2, y_2), (x_3, y_3)$.

Step 3: Write an equation in the form $y = mx + b$ for the line through (x_1, y_1) and (x_3, y_3).

Step 4: Use your values from Steps 2 and 3 to write an equation of the median-median line $y = mx + \frac{2}{3}b + \frac{1}{3}(y_2 - mx_2)$.

   1. $(-8, -28), (-3, -16), (-5, -18), (-1, -7), (5, 12),$
      $(0, -6), (3, 4), (7, 15), (2, 5)$

   2. $(-3, 16), (-1, 11), (1, 7), (0, 6), (2, 3), (5, 0), (4, -1), (6, -3), (9, -8),$

In Exercises 3–7, use the table which shows the approximate exchange rate of the Japanese yen per United States dollar in various years. Round decimals in equations to the nearest tenth.

Year	1970	1975	1980	1985	1990	1995
Number of yen per dollar	358	297	227	239	145	94

   3. Make a scatter plot of the data. Then use (0, 358) and (25, 94) to write a linear model for the exchange rate between the dollar and the yen *t* years after 1970.

   4. Why is the slope in the model from Exercise 3 negative? What does this mean for the strength of the yen against the dollar?

   5. Use the method from Exercises 1 and 2 to write a linear model for the data.

   6. Use each model from Exercises 3 and 5 to estimate the exchange rate in 2000.

TEACHER'S NAME _____ CLASS _____ ROOM _____ DATE _____

Lesson Plan

2-day lesson (See *Pacing the Chapter,* TE page 266A)   For use with pages 305–312

GOAL **Write equations of perpendicular lines.**

State/Local Objectives _____

_____

✓ Check the items you wish to use for this lesson.

STARTING OPTIONS
_____ Homework Check (5.5): TE page 301; Answer Transparencies
_____ Homework Quiz (5.5): TE page 304, CRB page 69, or Transparencies
_____ Warm-Up: CRB page 69 or Transparencies

TEACHING OPTIONS
_____ Developing Concepts: SE page 305; CRB page 70 (Activity Support Master)
_____ Lesson Opener: CRB page 71 or Transparencies
_____ Examples: Day 1: 1–2, SE pages 306–307; Day 2: 3, SE page 308
_____ Extra Examples: TE pages 307–308 or Transparencies; Internet Help at *www.mcdougallittell.com*
_____ Checkpoint Exercises: Day 1: Exs. 1–4, SE pages 306–307; Day 2: Ex. 5, SE page 308
_____ Graphing Calculator Activity with Keystrokes: CRB pages 72–73
_____ Concept Check: TE page 308
_____ Guided Practice Exercises: SE page 309; Day 1: Exs. 1–8; Day 2: Exs. 9–10

APPLY/HOMEWORK
Homework Assignment
_____ Transitional: Day 1: pp. 309–312, Exs. 11, 12, 17, 20, 21, 40, 41, 49–69 odd;
Day 2: pp. 310–312, Exs. 26, 27, 32, 33, 45, 47, 48, Quiz 3
_____ Average: Day 1: pp. 309–312, Exs. 13, 14, 18, 22, 23, 42, 43, 50–70 even;
Day 2: pp. 310–312, Exs. 28, 29, 34, 35, 44, 47, 48, Quiz 3
_____ Advanced: Day 1: pp. 309–312, Exs. 15, 16, 19, 24, 25, 52–58, 63–70;
Day 2: pp. 310–312, Exs. 30, 31, 36–39, 44–50*, Quiz 3; EC: CRB p. 80

Reteaching the Lesson
_____ Practice Masters: CRB pages 74–75 (Level A, Level B)
_____ Reteaching with Practice: CRB pages 76–77 or Practice Workbook with Examples;
Resources in Spanish
_____ Personal Student Tutor: CD-ROM

Extending the Lesson
_____ Interdisciplinary/Real-Life Applications: CRB page 79
_____ Challenge: CRB page 80

ASSESSMENT OPTIONS
_____ Daily Quiz (5.6): TE page 311 or Transparencies
_____ Standardized Test Practice: SE page 311; STP Workbook; Transparencies
_____ Quiz 5.5–5.6: SE page 312

Notes _____

_____

_____

TEACHER'S NAME _____ CLASS _____ ROOM _____ DATE _____

Lesson Plan for Block Scheduling

1-block lesson (See *Pacing the Chapter,* TE page 266A)   **For use with pages 305–312**

GOAL   Write equations of perpendicular lines.

State/Local Objectives _____
_____
_____

✓ **Check the items you wish to use for this lesson.**

STARTING OPTIONS

____ Homework Check (5.5): TE page 301; Answer Transparencies
____ Homework Quiz (5.5): TE page 304,
         CRB page 69, or Transparencies
____ Warm-Up: CRB page 69 or Transparencies

TEACHING OPTIONS

____ Developing Concepts: SE page 305; CRB page 70 (Activity Support Master)
____ Lesson Opener: CRB page 71 or Transparencies
____ Examples: 1–3, SE pages 306–308
____ Extra Examples: TE pages 307–308 or Transparencies; Internet Help at *www.mcdougallittell.com*
____ Checkpoint Exercises: Exs. 1–5, SE pages 306–308
____ Graphing Calculator Activity with Keystrokes: CRB pages 72–73
____ Concept Check: TE page 308
____ Guided Practice Exercises: SE page 309; Exs. 1–10

APPLY/HOMEWORK

Homework Assignment

____ Block Schedule: pp. 309–312: Exs. 13, 14, 18, 22, 23, 28, 29, 34, 35, 42, 43, 44, 47, 48, 50–70 even, Quiz 3

Reteaching the Lesson

____ Practice Masters: CRB pages 74–75 (Level A, Level B)
____ Reteaching with Practice: CRB pages 76–77 or Practice Workbook with Examples; Resources in Spanish
____ Personal Student Tutor: CD-ROM

Extending the Lesson

____ Interdisciplinary/Real-Life Applications: CRB page 79
____ Challenge: CRB page 80

ASSESSMENT OPTIONS

____ Daily Quiz (5.6): TE page 311 or Transparencies
____ Standardized Test Practice: SE page 311; STP Workbook; Transparencies
____ Quiz 5.5–5.6: SE page 312

Notes _____
_____
_____

CHAPTER PACING GUIDE	
Day	Lesson
1	5.1 (all); 5.2 (begin)
2	5.2 (end); 5.3 (all)
3	5.4 (all); 5.5 (all)
4	**5.6 (all)**
5	Ch. 5 Review and Assess

NAME _____ DATE _____

WARM-UP EXERCISES

For use before Lesson 5.6, pages 305–312

Write in standard form an equation of the line.

1.

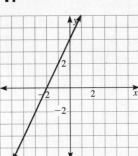

2.

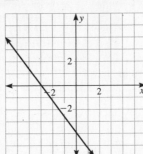

3.

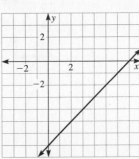

Write in standard form an equation of the line that passes through the two points. Use integer coefficients.

4. $(3, -2), (4, 5)$ **5.** $(6, 3), (-2, 1)$ **6.** $(0, -5,), (3, -2)$

DAILY HOMEWORK QUIZ

For use after Lesson 5.5, pages 298–304

You release a helium balloon attached to 500 feet of string on a windless day. The balloon climbs upward at a rate of 7 feet per second. You release the balloon at time $t = 0$.

1. What is the slope of the linear model for this situation?

2. The y-intercept represents the height of the balloon above the ground when you release it. What is the y-intercept of the line?

3. Use the slope and the y-intercept to write a linear model for the height y (in feet) of the balloon above the ground in terms of time t (in seconds). Use slope-intercept form.

4. After 30 seconds, have you run out of string?

5. Use the equation from Problem 3 to determine the time that it will take to use 490 feet of string.

Lesson 5.6

Lesson 5.6

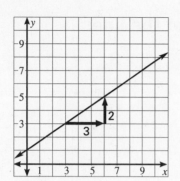

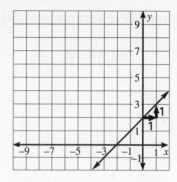

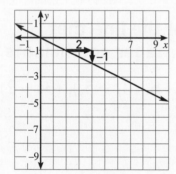

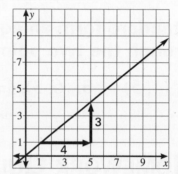

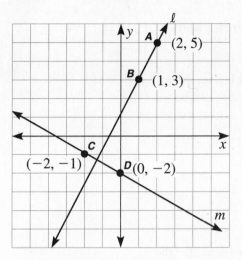

In the graph to the left, lines ℓ and m are perpendicular to each other.

1. Find the slope of each line using the coordinates of A and B on line ℓ and C and D on line m.

2. Multiply the slopes of the lines. What is the product?

Lesson 5.6

If two lines are perpendicular to each other, then the product of their slopes is –1.

3. What is the slope of a line perpendicular to a line whose slope is 3?

4. What is the slope of a line perpendicular to a line whose slope is $\frac{1}{5}$?

NAME _____ DATE _____

Graphing Calculator Activity

For use with pages 305–312

GOAL **To determine whether two different lines in the same plane are perpendicular.**

Geometrically, perpendicular lines intersect to form a right angle. Algebraically, the slopes of perpendicular lines also relate. A graphing calculator can be used to visually check whether two lines are perpendicular and to discover this slope relationship.

Activity

1 Enter the following equations into your graphing calculator.

$$y = \frac{4}{5}x + 2 \qquad\qquad y = -\frac{4}{5}x + 2$$

2 Plot the graph of each equation in the same coordinate plane. Do the lines appear perpendicular?

3 Compare the slopes of the equations.

4 Repeat Steps 1–3 with each pair of equations. Be sure to clear out the equations from Step 1.

   a. $y = \frac{4}{5}x + 2,\ y = \frac{5}{4}x + 2$ **b.** $y = \frac{4}{5}x + 2,\ y = -\frac{5}{4}x + 2$

5 Write a statement about the slopes of perpendicular lines.

Exercises

1. Determine the unknown slope that would make the lines perpendicular. Then use your graphing calculator to visually check your answer.

   a. $y = \frac{1}{3}x$ **b.** $y = -2x + 3$ **c.** $y = \frac{3}{2}x - 5$

   $y = \underline{\ ?\ }x$ $y = \underline{\ ?\ }x$ $y = \underline{\ ?\ }x$

2. Start with the equation $y = \frac{1}{2}x$. Write three other linear equations that, along with the first equation, will create a rectangle. Use your graphing calculator to visually check your answer.

See page 00 for keystrokes.

NAME _____ DATE _____

Graphing Calculator Activity

For use with pages 305–312

TI-82

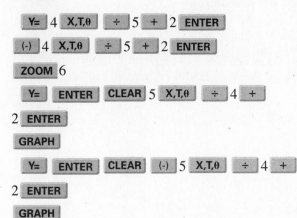

TI-83

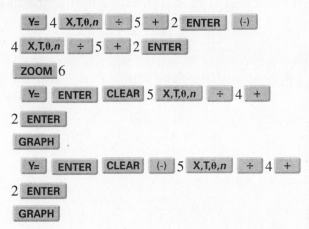

SHARP EL-9600c

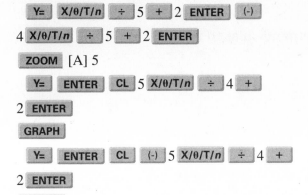

CASIO CFX-9850GA PLUS

From the main menu, choose GRAPH.

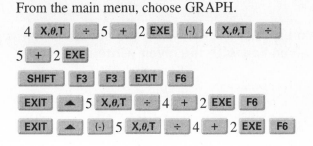

NAME _____ DATE _____

Practice A

For use with pages 305–312

In Exercises 1–10, determine whether the lines are perpendicular.

1. $y = 3x - 8, y = -\frac{1}{3}x + 2$

2. $y = \frac{1}{2}x + 3, y = 2x - 1$

3. $y = \frac{1}{4}x + 6, y = -4x - 6$

4. $y = \frac{2}{3}x - 1, y = -\frac{3}{2}x + 10$

5. $y = 8x - 3, y = \frac{1}{8}x + 3$

6. $y = x + 4, y = -x + 4$

7. $y = -\frac{3}{8}x + 2, y = \frac{8}{3}x - 3$

8. $y = -\frac{1}{5}x - 5, y = 5x + 2$

9. $y = \frac{3}{5}x - 9, y = \frac{5}{3}x - 1$

10. $y = \frac{2}{7}x + 9, y = -\frac{7}{2}x - 4$

In Exercises 11–14, write the equation of the line passing through the two points. Show that this line is perpendicular to the given line.

11. $(4, 2)$ and $(-4, 6)$; $y = 2x + 3$

12. $(-3, 8)$ and $(0, 11)$; $y = -x - 6$

13. $(5, -6)$ and $(3, 2)$; $y = \frac{1}{4}x + 4$

14. $(2, 0)$ and $(8, -3)$; $y = 2x$

In Exercises 15–18, write the equation that is perpendicular to the given line with the given y-intercept.

15. $y = 2x + 6, b = 3$

16. $y = -\frac{3}{8} - 4, b = -2$

17. $y = \frac{2}{3}x + 3, b = 5$

18. $y = -5x - 3, b = -7$

In Exercises 19–21, write the equation of each line in the graph. Determine whether the lines are perpendicular.

19.
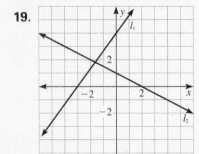

20.
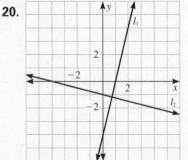

21.
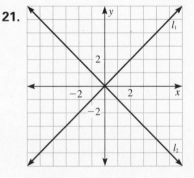

NAME _____ DATE _____

Practice B

For use with pages 305–312

In Exercises 1–6, determine whether the lines are perpendicular.

1. $y = \frac{1}{2}x - 3, y = -2x + 6$

2. $y = -10x + 1, y = \frac{1}{10}x + 1$

3. $y = \frac{2}{3}x + 7, y = \frac{3}{2}x - 8$

4. $y = 4x - 6, y = -4x - 5$

5. $y = 2x, y = -2x$

6. $y = \frac{4}{3}x - 3, y = \frac{-3}{4}x$

In Exercises 7–10, write the equation of the line passing through the two points. Show that this line is perpendicular to the given line.

7. $(-1, 10)$ and $(-8, -4)$; $y = -\frac{1}{2}x + 3$

8. $(4, 1)$ and $(6, -5)$; $y = \frac{1}{3}x - 7$

9. $(9, -7)$ and $(3, 11)$; $y = \frac{1}{3}x + 6$

10. $(1, -6)$ and $(3, -10)$; $y = \frac{1}{2}x - 5$

In Exercises 11–18, write the equation that is perpendicular to the given line with the given y-intercept.

11. $y = \frac{2}{3}x - 8, b = 6$

12. $y = \frac{1}{2}x + 6, b = -3$

13. $y = 6x + 7, b = -1$

14. $y = x + 1, b = -1$

15. $y = -5x - 5, b = 10$

16. $y = -\frac{3}{4}x - 3, b = \frac{1}{2}$

17. $y = \frac{5}{3}x + \frac{3}{5}, b = \frac{3}{4}$

18. $y = -\frac{3}{2}x - 1, b = -8$

In Exercises 19–22, write in slope-intercept form the equation of the line passing through the given point and perpendicular to the given line.

19.

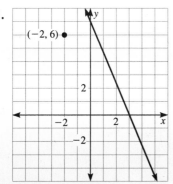

20.

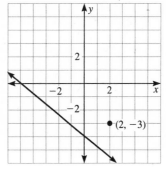

21.

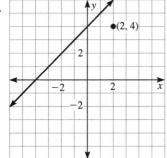

22.

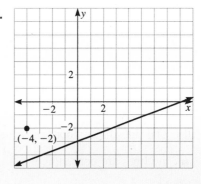

NAME _____ DATE _____

Reteaching with Practice

For use with pages 305–312

GOAL **Write equations of perpendicular lines.**

EXAMPLE 1 *Identify Perpendicular Lines*

Determine whether the lines are perpendicular.

SOLUTION

The lines have slopes of 1 and -1.
Because $(1)(-1) = -1$, the lines
are perpendicular.

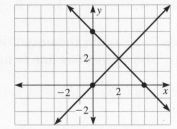

Exercises for Example 1

Determine whether the lines are perpendicular.

1. $y = -4x + 7, y = \frac{1}{4}x - 3$

2. $y = -\frac{2}{3}x + 3, y = -\frac{3}{2}x - 2$

EXAMPLE 2 *Show that Lines are Perpendicular*

a. Write in slope-intercept form the equation of the line passing through
$(-2, 1)$ and $(6, -2)$.

b. Show that the line is perpendicular to the line $y = \frac{8}{3}x - 2$.

SOLUTION

a. 1. Find the slope. Let $(x_1, y_1) = (-2, 1)$ and $(x_2, y_2) = (6, -2)$.

$$m = \frac{y_2 - y_1}{x_2 - x_1} = \frac{-2 - 1}{6 - (-2)} = \frac{-3}{8} = -\frac{3}{8}$$

2. Write the equation of the line using point-slope form.

$y - y_1 = m(x - x_1)$ Write point-slope form.

$y - 1 = -\frac{3}{8}(x + 2)$ Substitute $-\frac{3}{8}$ for m, -2 for x_1, and 1 for y_1.

$y - 1 = -\frac{3}{8}x - \frac{3}{4}$ Use distributive property.

$y = -\frac{3}{8}x + \frac{1}{4}$ Add 1 to each side.

b. The lines have slopes of $-\frac{3}{8}$ and $\frac{8}{3}$. Because $\left(-\frac{3}{8}\right)\left(\frac{8}{3}\right) = -1$,

the lines are perpendicular.

Reteaching with Practice

For use with pages 305–312

Exercises for Example 2

3. Write in slope-intercept form the equation of the line passing through $(1, -3)$, and $(-2, -2)$. Show that the line is perpendicular to $y = 3x - 4$.

4. Write in slope-intercept form the equation of the line passing through $(-4, 6)$ and $(2, -3)$. Show that the line is perpendicular to $y = \dfrac{2}{3}x + 5$.

EXAMPLE 3 *Writing Equations of Perpendicular Lines*

Write an equation of the line that is perpendicular to the line $y = -3x + 2$ and passes through the point $(6, 5)$.

SOLUTION

The given line has a slope of $m = -3$. A perpendicular line through $(6, 5)$ must have a slope of $m = \dfrac{1}{3}$. Use this information to find the y-intercept.

$y - y_1 = m(x - x_1)$ Write point-slope form.

$y - 5 = \dfrac{1}{3}(x - 6)$ Substitute $\dfrac{1}{3}$ for m, 6 for x_1, and 5 for y_1.

$y - 5 = \dfrac{1}{3}x - 2$ Use distributive property.

$y = \dfrac{1}{3}x + 3$ Add 5 to each side.

Exercises for Example 3

Write an equation of the line that is perpendicular to the given line and passes through the given point.

5. $y = 2x - 1$, $(2, 4)$ **6.** $y = -\dfrac{1}{3}x + 2$, $(5, 1)$ **7.** $y = -4x + 5$, $(4, 3)$

Lesson 5.6

NAME _____ DATE _____

Quick Catch-Up for Absent Students

For use with pages 305–312

The items checked below were covered in class on (date missed) _____

Developing Concepts Activity: Perpendicular Lines

_____ **Goal:** Describe the relationship between the slopes of perpendicular lines.

Lesson 5.6: Perpendicular Lines

_____ **Goal:** Write equations of perpendicular lines.

Material Covered:

_____ Student Help: More Examples

_____ Example 1: Identify Perpendicular Lines

_____ Student Help: Study Tip

_____ Example 2: Show that Lines are Perpendicular

_____ Example 3: Write an Equation of a Perpendicular Line

Vocabulary:

perpendicular, p. 306

_____ Other (specify) _____

_____

Homework and Additional Learning Support

_____ Textbook (specify) pp. 309–312 _____

_____ Internet: Extra Examples at www.mcdougalllittell.com

_____ *Reteaching with Practice* worksheet (specify exercises)_____

_____ *Personal Student Tutor* for Lesson 5.6

Interdisciplinary Application:
Perpendicular Lines

For use with pages 305–312

**There are three ways to construct a perpendicular
line. In this lesson we will explore two of those
ways. You will need a compass and straightedge.**

Construct the perpendicular bisector of a given segment.

   A. Place the metal tip of the compass on point *A*.

   B. Open the compass to a length greater than half the distance
   from *A* to *B*.

   C. Draw an arc from point *A* above and below
   the line.

   D. Retaining the distance between the compass points, draw a
   second arc from point *B* that intersects the first arc above and
   below the line.

   E. Using a straightedge, draw a line that connects the
   intersection of the two arcs. The resulting line divides segment
   AB into two equal segments and is perpendicular to line *AB*.

Construct the perpendicular line to a given line at a given point on the line.

   A. Place the metal tip of the compass on point *C*, open the
   compass to any length and draw arcs on either side of *C* on the
   line. Label the points where the arcs
   cross the line as *D* and *E*.

   B. Now use *D* and *E* as the points to place the metal tip and
   open the compass to a length greater than the distance from
   C to *D*. Draw an arc from each point so that the two arcs
   intersect above the line. Label this intersection point as *F*.

   C. Using a straightedge draw line *CF*. Line *CF* is perpendicular
   to the given line.

1. Follow the instructions below to complete the given construction.

   A. Draw a circle, and label the center *C*. (Use any size radius.)

   B. Draw diameter *AB*.

   C. Construct the perpendicular bisector of segment *AB*. Label the
   points where it intersects the circle as *D* and *E*.

   D. Construct the perpendicular bisector of segment *CB*, label it
   FG, and label as *H* the point intersecting segment *CB*.

   E. With *H* as the center and *DH* as the compass length, make an
   arc intersecting segment *AC*, labeling the point of intersection
   as *J*.

   F. With *D* as the center and *DJ* as the compass length, make an
   arc intersecting the circle, labeling that point *K*.

   G. Starting with *K* as the center and *DK* as the compass length,
   mark successive arcs around the circle, labeling the points as
   as *L*, *M*, and *N*.

   H. Using a straightedge draw segments *DK*, *KL*, *LM*, *MN* and *ND*.

   I. What figure have you just constructed?

NAME _____ DATE _____

Challenge: Skills and Applications

For use with pages 305–312

In Exercises 1–6, determine whether the lines are perpendicular.

1. $y = 0.1x - 6; y = -10x + 3$
2. $y = 0.8x - 7; y = -1.25x + 6$
3. $y = 2.5x + 5; y = -\dfrac{1}{4}x + 2$
4. $3y = 2x - 7; y = -\dfrac{2}{3}x + 6$
5. $4y = -5x + 2; 5y = 4x - 7$
6. $3y = 4x - 8; y = -0.75x + 1$

In Exercises 7–10, write the equation that is perpendicular to the given line with the given y-intercept.

7. $y = 0.25x + 2, b = 6$
8. $y = -3.2x - 1, b = -\dfrac{1}{2}$
9. $y = -1.6x - 7, b = 0$
10. $y = 0.2x + 6, b = -8$

In Exercises 11–14, *ABCD* is a rectangle, and coordinates are given for *A, B,* and *C*. Find the coordinates of *D*.

11. $A(2, 4), B(6, 4), C(6, -5)$
12. $A(-6, 3), B(-4, 3), C(-4, -3)$
13. $A(-4, 4), B(4, 4), C(4, -1)$
14. $A(1, 3), B(4, 1), C(0, -5)$

Geometry Connection   **In Exercises 15–17, use the graph.**

15. Find the perpendicular sides of trapezoid *WXYZ*. How do you know mathematically that these two sides are perpendicular?

16. Write equations of the lines passing through the perpendicular sides.

17. Write equations of the lines passing through the two parallel sides. How do you know mathematically that these two sides are parallel?

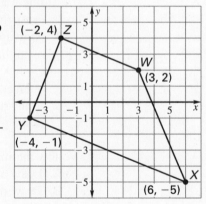

Chapter Review Games and Activities

For use after Chapter 5

Using the given information, write linear equations in SLOPE-INTERCEPT or STANDARD form. The letter associated with the equation in the correct form will answer the riddle when placed on the line with the problem number.

Which trees in the forest get invited to the most parties?

Given information

1. $m = -\dfrac{6}{5}$ $b = -2$

6. $(0, 6)$ $m = \dfrac{6}{7}$

    (S) $x + 3y = 9$

    (T) $y = \dfrac{6}{5}x + 2$

    (A) $x + 2y = 6$

    (P) $3x - 7y = 49$

2. $m = \dfrac{1}{2}$ $b = 3$

7. $(-7, 7)$ $(9, -3)$

    (S) $8x - 7y = -42$

    (P) $6x + 5y = -10$

    (O) $y = \dfrac{1}{2}x + 3$

    (R) $6x - 7y = -42$

3. $m = \dfrac{3}{7}$ $b = -7$

8. $(1, -1)$ $(4, 5)$

    (N) $2x - y = 3$

    (E) $y = \dfrac{1}{2}x - 1$

    (L) $8x - 7y = 70$

    (O) $y = \dfrac{6}{7}x - 6$

4. $(7, -2)$ $m = \dfrac{8}{7}$

9. $(-6, -4)$ $(-2, -2)$

    (W) $x + 2y = 2$

    (A) $y = -\dfrac{13}{6}x - 7\dfrac{2}{3}$

    (O) $y = -\dfrac{5}{8}x + 2\dfrac{5}{8}$

5. $(-4, 1)$ $m = -\dfrac{13}{6}$

10. $(9, 0)$ $(-3, 4)$

___ ___ ___ ___ ___ ___   ___ ___ ___ ___
(1) (2) (3) (4) (5) (6)   (7) (8) (9) (10)

Review and Assess

Chapter Test A

For use after Chapter 5

In Exercises 1 and 2, write an equation of the line in slope-intercept form.

1. The slope is -5; the y-intercept is 7.

2. The slope is 10; the y-intercept is -3.

Write an equation in slope-intercept form of the line shown in the graph.

3.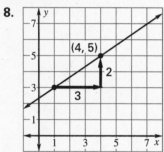

4.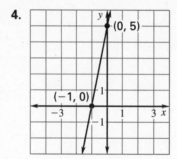

5. Write a linear equation to model the situation. You borrow $70 from your brother. To repay the loan, you pay him $7 per week.

Write an equation of the line that passes through the point and has the given slope. Write the equation in slope-intercept form.

6. $(3, 0), m = -2$

7. $(1, 2), m = 2$

Write an equation in slope-intercept form of the line shown in the graph.

8.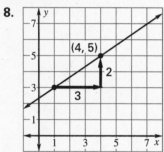

9.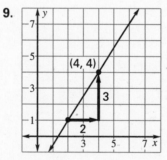

Write an equation of the line that is parallel to the given line and passes through the given point.

10. $y = x + 3, (5, 0)$

11. $y = 2x + 3, (-4, 1)$

Answers

1. _____

2. _____

3. _____

4. _____

5. _____

6. _____

7. _____

8. _____

9. _____

10. _____

11. _____

Write an equation in slope-intercept form of the line that passes through the points.

12. $(-4, 2), (1, -1)$

13. $(-2, -1), (3, 5)$

14. Write an equation of a line that is perpendicular to $y = 2x + 3$ and passes through $(3, 4)$.

Write an equation in point-slope form of the line that passes through the given points.

15. $(-3, -4), (3, 4)$

16. $(-5, -4), (7, -5)$

Write the equation in standard form with integer coefficients.

17. $5x - y + 6 = 0$

18. $y = -3x + 9$

Write the equations in standard form of the horizontal and vertical lines.

19.

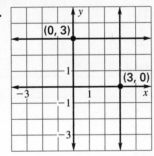

20.

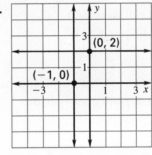

21.

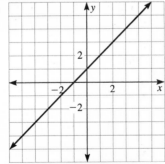

Write the equation of a line that is parallel to the given line and passes through the point $(2, -1)$.

22.

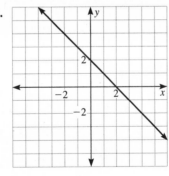

Write an equation of a line that is perpendicular to the given line and passes through the point $(1, 3)$.

Answers
12._____
13._____
14._____
15._____
16._____
17._____
18._____
19._____
20._____
21._____
22._____

Review and Assess

NAME _____ DATE _____

Chapter Test B

For use after Chapter 5

In Exercises 1 and 2, write an equation of the line in slope-intercept form.

1. The slope is -3; the y-intercept is 5.

2. The slope is 4; the y-intercept is 0.

Write an equation in slope-intercept form of the line shown in the graph.

3.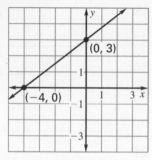

4.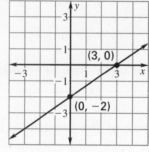

5. Write a linear equation to model the situation. You walked 4 miles on a trail. You continue to walk at a rate of 3 miles per hour.

Write an equation of the line that passes through the point and has the given slope. Write the equation in slope-intercept form.

6. $(3, 2)$, $m = \dfrac{1}{2}$

7. $(-3, 2)$, $m = \dfrac{1}{2}$

Write an equation in slope-intercept form of the line shown in the graph.

8.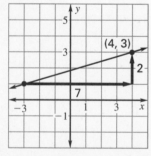

9.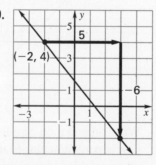

Write an equation of the line that is parallel to the given line and passes through the given point.

10. $y = -3x + 2$, $(2, 3)$

11. $y = \dfrac{1}{2}x - 5$, $(-3, -1)$

Write an equation in slope-intercept form of the line that passes through the points.

12. $(-5, 3), (4, -5)$

13. $\left(-\dfrac{1}{2}, -1\right), \left(3, \dfrac{5}{2}\right)$

Answers

1. _____

2. _____

3. _____

4. _____

5. _____

6. _____

7. _____

8. _____

9. _____

10. _____

11. _____

12. _____

13. _____

Review and Assess

NAME _____ DATE _____

Chapter Test B

For use after Chapter 5

14. Write an equation of a line that is perpendicular to $y = -3x + 5$ and passes through $(4, 3)$.

Write an equation in point-slope form of the line that passes through the given points.

15. $(5, -6), (1, -7)$ **16.** $(6, -3), (-1, 9)$

Write the equation in standard form with integer coefficients.

17. $0.5x - 2y - 0.75 = 0$ **18.** $y = -\dfrac{1}{3}x - 5$

Write the equations in standard form of the horizontal and vertical lines that pass through the point.

19. $(2, 4)$ **20.** $(-5, 4)$

21. You are at the music store looking for CDs. The store has CDs for $10 and $15. You have $55 to spend. Write an equation that represents the different numbers of $10 and $15 CDs that you can buy.

Answers
14. _____
15. _____
16. _____
17. _____
18. _____
19. _____
20. _____
21. _____
22. _____
23. _____

22.

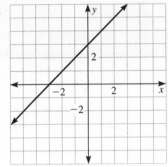

Write an equation of a line that is parallel to the given line and passes through the point $(0, 0)$.

23.

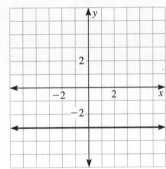

Write an equation of a line that is perpendicular to the given line and passes through the point $(0, 2)$.

Review and Assess

NAME _____ DATE _____

SAT/ACT Chapter Test

For use after Chapter 5

1. What is the equation of the line that passes through the points $(-3, 4)$ and $(-9, 6)$?

   A $y = -\frac{1}{3}x - \frac{5}{3}$ **B** $y = -\frac{1}{3}x + 3$

   C $y = -3x - 5$ **D** $y = -3x + 12$

2. A line with a slope of -3 passes through the point $(4, -3)$. What is the equation of the line in standard form?

   A $y = 3x + 9$ **B** $y = -3x + 9$

   C $3x + y = 9$ **D** $y + 3 = -3(x - 4)$

3. What is the equation of the line that is parallel to the line $y = \frac{1}{3}x - 2$ and passes through $(3, -5)$?

   A $y = -3x + 4$ **B** $y = \frac{1}{3}x + \frac{14}{3}$

   C $y = -3x - 12$ **D** $y = \frac{1}{3}x - 6$

4. What is the equation of the line that is perpendicular to the line $y = -\frac{3}{4}x + 4$ and has a y-intercept of -5?

   A $y = -\frac{3}{4}x - 5$ **B** $y = \frac{3}{4}x - 5$

   C $y = \frac{4}{3}x - 5$ **D** $y = -\frac{4}{3}x + 5$

5. What is the equation of the line that passes through $(-6, 2)$ and has a slope of $-\frac{2}{3}$?

   A $y = -\frac{2}{3}x - \frac{14}{3}$ **B** $y = -\frac{2}{3}x - 2$

   C $y = -\frac{2}{3}x + 6$ **D** $y = -\frac{2}{3}x - 6$

6. What is the equation of the line shown in the graph?

   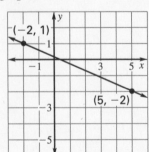

   A $y = -x - 1$

   B $y = -\frac{3}{7}x + 3$

   C $y = -x + 3$

   D $y = -\frac{3}{7}x + \frac{1}{7}$

In Exercises 7 and 8, choose the statement below that is true about the given numbers.

   A. The number in column A is greater.
   B. The number in column B is greater.
   C. The two numbers are equal.
   D. The relationship cannot be determined from the given information.

7.

Column A	Column B
x-intercept of $2x - 3y = 4$	x-intercept of $3x - 7y = -6$

   A **B** **C** **D**

8.

Column A	Column B
slope of $9x - 12y = 8$	slope of $4y - 3x = 16$

   A **B** **C** **D**

9. What is the equation of the line that is perpendicular to the line shown in the graph?

   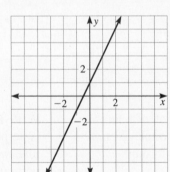

   A $y = -\frac{1}{2}x + 1$

   B $y = -2x - 1$

   C $y = \frac{1}{2}x + 1$

   D $y = 2x - 1$

Alternative Assessment and Math Journal

For use after Chapter 5

JOURNAL **1.** In Chapter Five we have been studying writing equations of lines. In this journal we will explore linear equations in the real world. (a) Create a situation in which you can model change with a linear equation. Describe the situation in your journal and give at least two data points. (b) Write a linear equation based on the situation you described. Use slope-intercept or point-slope form. (c) Graph the linear equation that you have created. (d) Predict a point on the line. Check your prediction with the linear equation.

MULTI-STEP PROBLEM **2.** Zachary and Jeremy both have savings accounts at the local bank. Jeremy has a job and has been saving his money. Zachary relies on his account for spending money.

- Jeremy had $500 in his account at the end of 3 weeks, and $800 in his account at the end of 7 weeks.

- Zachary withdraws $7.50 per week from his account. After twelve weeks, the balance in his account was $432.50.

a. Find the rate of change in both Zachary and Jeremy's accounts. Describe what each means in terms of the particular account.

b. Write a model for each which gives the balance of each account, y, in terms of the number of weeks, x.

c. Sketch a graph of each model and label your axes.

d. How much money will be in each account after 20 weeks?

e. After how many weeks will Zachary run out of money? Can he make a withdrawal of $7.50 in the final week? If yes, explain. If no, how much can he withdraw?

3. *Critical Thinking* Predict when Jeremy and Zachary will have the same amount in their savings accounts. Show three different methods that can be used to solve this problem. Show all work and explain why you can solve the problem using these methods. Explain why it can be helpful to have more than one method for solving a particular problem.

Review and Assess

Alternative Assessment and Math Journal

For use after Chapter 5

JOURNAL SOLUTION

1. a–d: Complete answers should address these points.

- Explain that linear models are linear functions that are used to model a real-life situation. The rate of change is a method of comparing the change of two quantities.

  a. The situation should be able to be modeled with a linear equation. It should involve a rate of change. At least two data points should be given.

  b. A linear equation based on the points chosen in part (a) should be given in slope-intercept or point-slope form.

  c. Check graphs. The graph should be linear.

  d. The point should fall on the line or the student should have explained why the prediction was not accurate.

MULTI-STEP PROBLEM SOLUTION

2. a. 75; -7.5; Jeremy deposits $75 weekly; Zachary withdraws $7.50 weekly.

b. Jeremy: $y = 75x + 275$; Zachary: $y = -7.50x + 522.50$

c. Check graphs.

d. Jeremy: $1775; Zachary: $372.50

e. 70 weeks; He can withdraw $5 in the last week.

3. *Critical Thinking Solution* They would be the same after 3 weeks. *Sample answer:* One method would be to set the two equations equal to each other, $75x + 275 = -7.50x + 522.50$, and solve for x. Another method would be to sketch the graph of both equations and find out where the two equations have the same x-value. This would occur where the two equations intersect. A third method would be to use a table of values, either on the calculator or by hand, and calculate the balances for several weeks to determine the solution.

MULTI-STEP PROBLEM RUBRIC

4 Students complete all parts of the questions accurately. Explanations are logical. Three different methods are clearly explained for determining when the accounts would be equal. Equations and graphs are completely correct.

3 Students complete the questions and explanations. Solutions may contain minor mathematical errors, but errors are carried through (i.e. if the incorrect equations are given, these incorrect equations are graphed accurately). Graphs are correct. Some explanations may not be completely clear.

2 Students complete questions and explanations. Several mathematical errors may occur. Explanations are not logical. Graph is incomplete or incorrect.

1 Answers are very incomplete. Solutions and reasoning are incorrect. Graph is missing or is completely inaccurate. Explanations are not logical.

Review and Assess

Project: What We Read

For use with Chapter 5

OBJECTIVE Analyze the amounts spent by United States consumers on books and maps compared with magazines, newspapers, and sheet music.

MATERIALS paper, pencil, calculator, graph paper

INVESTIGATION The Bureau of Economic Analysis, United States Department of Commerce, groups reading materials as shown in the table.

Personal Consumer Purchases in the United States
(in billions of dollars)

	1990	1996
Books and maps	17.5	23.2
Magazines, newspapers, and sheet music	23.8	26.5

1. Use the datapoints in the table to write a linear model for the amount spent on books and maps. Use the datapoints in the table to write a linear model for the amount spent on magazines, newspapers, and sheet music. For each, let t be years since 1990 and y be amount spent.

2. Use the linear models from Exercise 1 to estimate the amount of purchases of each type in 1993.

3. In 1993, people in the United States spent $22.1 billion on books and maps and $22.7 billion on magazines, newspapers, and sheet music. Which of your estimates from Exercise 2 was closer to the actual amount?

4. Use the linear models from Exercise 1 to estimate the amount of purchases of each type in 2000.

5. Find the amount spent on each type of reading material in the year 2000 according to the Bureau of Economic Analysis. Which of your estimates from Exercise 4 was closer to the actual amount?

PRESENT YOUR RESULTS Analyze which linear model fits the data better overall. Write a report about your analysis. Include your equations, estimates, and predictions.

Review and Assess

Project: Teacher's Notes

For use with Chapter 5

GOALS • Write an equation of a line given two points on the line.

• Use a linear equation to model a real-life problem.

• Use a linear model to make a real-life prediction.

MANAGING THE PROJECT You may wish to have students write the equations and make predictions before doing research in Exercise 5.

RUBRIC The following rubric can be used to assess student work.

4 The student finds the linear models for the amount spent on books and maps and for the amount spent on magazines, newspapers, and sheet music, and uses the models to estimate and predict. The report demonstrates clear thinking and explanation. All work is complete and correct.

3 The student finds the linear models for the amount spent on books and maps and for the amount spent on magazines, newspapers, and sheet music, and uses the models to estimate and predict. However, the report indicates some minor misunderstanding of content, there are errors in computation, or the presentation is weak.

2 The student partially achieves the mathematical and project goals of finding the linear models for the amount spent on books and maps and for the amount spent on magazines, newspapers, and sheet music, and uses the models to estimate and predict. However, the report indicates a limited grasp of the main ideas or requirements. Some of the work is incomplete, misdirected, or unclear.

1 The student makes little progress toward accomplishing the goals of the project because of lack of understanding or lack of effort.

Cumulative Review

For use after Chapters 1-5

Evaluate the expression for the given values of the variable. (1.2)

1. $(2s + t)^3$ when $s = -1$ and $t = -5$

2. $a - \frac{1}{2}b^3$ when $a = \frac{1}{3}$ and $b = -\frac{2}{3}$

3. $-10.9(k - 0.36)$ when $k = 0.45$

4. $8 - (x^2) - (y^3)$ when $x = -3$ and $y = 6$

Write the sentence as an equation or an inequality. (1.5)

5. The length l of a table is three times its width w.

6. The time t it takes to go to work from home is less than one half the time s it takes to go to the mall.

7. The height h of a triangle is equal to the quotient of two times the area a and its base b.

Find the sum or difference. (2.3–2.4)

8. $3 - 2^2 - 2$

9. $-23.8 - 0.3 + 45.9$

10. $3.6 + 4.2 - 5.6$

11. $-\frac{7}{3} - \frac{5}{6} + \left(\frac{2}{3}\right)^2$

Solve the equation. Round to the nearest hundredth. (3.1–3.6)

12. $-14x - 5 = 93$

13. $-15a + 30 = -90$

14. $-(x - 2.3) - 4(5.9 - x) = 80$

15. $\frac{2}{9}\left(x - \frac{6}{7}\right) = \frac{1}{63}x + \frac{40}{63}$

Find the *x*- and *y*-intercepts of the equation. (4.4)

16. $y = 4x - 14$

17. $8x + 40 = 10y$

Find the slope of the line passing through the given points. (4.5)

18. $(0, -18), (-3, 18)$

19. $(8, 4), (50, 6)$

Write an equation of the line with the given slope and *y*-intercept. (5.1)

20. $m = 3, b = -2$

21. $m = -5, b = 4$

22. $m = 7, b = 11$

23. $m = \frac{1}{5}, b = -6$

24. $m = 0, b = 8$

25. $m = -6.5, b = 4.5$

Write an equation in slope-intercept form of the line that passes through the given points. (5.3)

26. $(12, -3), (-8, 1)$

27. $(-12, -56), (-40, 0)$

28. $(4, -4), (1, 5)$

29. $(-3, 3), (3, -3)$

Write an equation in point-slope form of the line that passes through the given points. (5.2, 5.3)

30. $(2, 7), (-2, -7)$

31. $(-10, 8), (-20, -12)$

32. $(4, 1), (-2, -3)$

33. $(-10, 10), (-6, -6)$

Review and Assess

Algebra 1
Chapter 5 Resource Book

Write an equation of the line that is parallel to the given line and passes through the given point. (5.2)

34. $y = 4x + 6$, $(4, -2)$

35. $y = -\frac{1}{4}x - 1$, $(4, 1)$

36. $y = -3x + 9$, $(3, -2)$

37. $y = \frac{4}{3}x - 6$, $(3, 1)$

Write the standard form of the equation of the line passing through the given point that has the given slope. (5.4)

38. $(2, -5)$, $m = -5$

39. $(0, 5)$, $m = -\frac{1}{2}$

40. $(-2, 7.5)$, $m = -6$

41. $(-4, 4)$, $m = 4$

In Exercises 42–44, use the following information regarding the farm population (in millions of persons) from 1945 to 1990. (5.5)

Year	1945	1990
Millions of Persons	24.3	4.4

42. Write a linear model for the farm population, y, in millions of persons. Let $x = 0$ represent 1945.

43. Use the linear model to estimate the average farming population in 1975.

44. Use the linear model to estimate the average farming population in 1980.

For Exercises 45-47, write in slope intercept form the equation of the line passing through the given point and perpendicular to the given line. (5.6)

45.
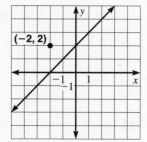

46.
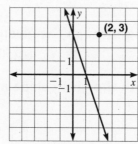

47.
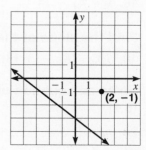

Algebra 1
Chapter 5 Resource Book

Review and Assess

ANSWERS

Chapter Support

Parent Guide
Chapter 5

5.1: $y = -\dfrac{1}{2}x - 3$ **5.2:** $y - 7 = \dfrac{1}{2}(x + 4)$

5.3: $y = -2x + 5$ **5.4:** $y - 4x = 16$

5.5: $10x + 15y = 60$; **5.6:** $y = -2x + 1$

Prerequisite Skills Review

1. -2 **2.** no solution **3.** 5 **4.** $\dfrac{1}{2}$

5.

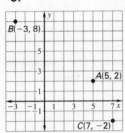

6.

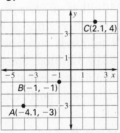

7.

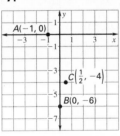

8.

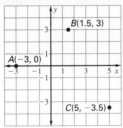

9. x-intercept $= 5$
y-intercept $= 25$

10. x-intercept $= 11$
y-intercept $= 3$

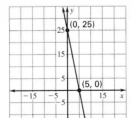

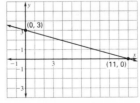

11. x-intercept $= -7$
y-intercept $= -\dfrac{35}{6}$

12. x-intercept $= 10$
y-intercept $= \dfrac{5}{4}$

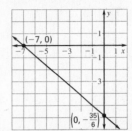

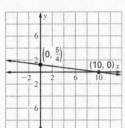

Strategies for Reading Mathematics

1. $635; $1035; $200 per week **2.** $35

3. $y = 225x + 40$

4. Total earnings = Earnings per extra chore × Number of extra chores + Allowance;
Total earnings = y (dollars),
Earnings per extra chore = 1.50 (dollars),
Number of extra chores = x (chores),
Allowance = 3 (dollars); $y = 1.5x + 3$

Lesson 5.1

Warm-Up Exercises

1. $\dfrac{1}{3}$ **2.** 3 **3.** $-10, -7, -1$

4. 173, 205, 269

Daily Homework Quiz

1. Yes. Domain is 0, 4, 8, 12, 16. Range is 1, 2, 3, 4, 5. **2.** No. It does not pass the vertical line test.
3. $f(-4) = -14$ **4.** $d = 12t$

Lesson Opener

Allow 10 minutes.

1. a. Check graphs. **b.** 1 **c.** 2 **d.** The y-intercept is the constant term in the equation and the slope is the coefficient of x. **2. a.** Check graphs **b.** -3 **c.** 1 **d.** The y-intercept is the constant term in the equation and the slope is the coefficient of x. **3. a.** Check graphs. **b.** 2
c. $-\dfrac{1}{2}$ **d.** The y-intercept is the constant term in the equation and the slope is the coefficient of x.
4. slope $= 5$, y-intercept $= -3$

Lesson 5.1 *continued*

Practice A

1. $2; 5$ **2.** $-4; 1$ **3.** $1; -5$ **4.** $\frac{1}{2}; 0$ **5.** $2; 3$

6. $2; -\frac{3}{2}$ **7.** $y = x$ **8.** $y = -2x + 4$

9. $y = -3x - 5$ **10.** $y = 6x - 1$

11. $y = 9$ **12.** $y = -6x - 2$ **13.** $y = 2x - 8$

14. $y = -4x + 11$ **15.** $y = 5x + 5$

16. $y = -5x - 4$ **17.** $y = -\frac{3}{5}x + 3$

18. $y = \frac{8}{9}x - \frac{1}{2}$ **19.** $y = x + 1$ **20.** $y = -x$

21. $y = 2$ **22.** $y = -x + 4$

23. $y = -2x - 4$ **24.** $y = \frac{3}{2}x - 3$

Practice B

1. $y = 2x + 3$ **2.** $y = 5x$ **3.** $y = 4x - 3$

4. $y = -5x + 1$ **5.** $y = -3x - 2$

6. $y = -5$ **7.** $y = \frac{1}{2}x - 8$ **8.** $y = -\frac{3}{4}x + 9$

9. $y = -\frac{1}{5}x + 3$ **10.** $y = \frac{4}{5}x - 7$

11. $y = \frac{1}{3}x + \frac{2}{3}$ **12.** $y = -\frac{4}{3}x + \frac{7}{8}$

13. $y = x + 2$ **14.** $y = -x + 3$

15. $y = 2x + 4$ **16.** $y = x - 4$ **17.** $y = \frac{1}{2}x + 1$

18. $y = -\frac{3}{2}x + 3$

Reteaching with Practice

1. $y = -2x + 5$ **2.** $y = x - 4$ **3.** $y = 2$

4. $y = 3x + 6$ **5.** $y = \frac{1}{3}x + 1$ **6.** $y = 2x - 4$

7. $y = -\frac{3}{4}x - 3$ **8.** $y = -\frac{5}{2}x + 5$

Interdisciplinary Application

1. $y = 7x + 2500; y = 18x$

2. 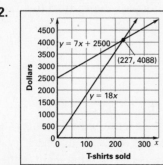 **3.** 227 T-shirts

4. $y = 0.55x + 50; y = 1x$

5. 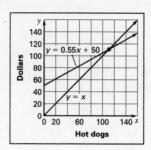 **6.** 111 hotdogs

Challenge: Skills and Applications

1. 2.45 million; the population of the United States increased an average of 2.45 million people per year between 1950 and 1990.

2. $y = 2.45t + 151$ **3.** 200 million; 3 million off; *Sample answer*: Yes, the approximation is off by only 1.5%. **4.** 298 million

5. $y = 23.36t + 285.7$ **6.** \$425.86 billion; \$5.24 billion off; *Sample answer*: It's a fairly close approximation since an error of 5.24 billion out of 431.1 billion is only about a 1% error.

7. \$519.3 billion; *Sample answer*: The prediction is probably somewhat close since the 1996 prediction was fairly close.

Lesson 5.2

Warm-Up Exercises

1. $-\frac{4}{7}$ **2.** $\frac{3}{7}$ **3.** $y = \frac{3}{5}x + 27$

4. $y = -2x + 5$

Daily Homework Quiz

1. $y = -2x + 3$ **2.** $-\frac{3}{2}; 4$ **3.** $y = 3x - 3$

4. $\frac{1}{3}; -5$ **5.** 1 mile

Lesson Opener

Allow 15 minutes.

1. a, b.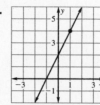

c. They are the same.

d. The 1 is subtracted from x and the 4 is subtracted from y.

e. It is multiplied by $(x - 1)$.

Answers

Lesson 5.2 *continued*

2. a, b.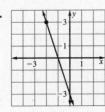

c. They are the same.

d. The -2 is subtracted from x and the 3 is subtracted from y.

e. It is multiplied by $(x + 2)$.

3. a, b.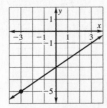

c. They are the same.

d. The -3 is subtracted from x and the -5 is subtracted from y.

e. It is multiplied by $(x + 3)$.

Practice A

1. $y - 5 = 2(x - 1)$ or $y + 1 = 2(x + 2)$
2. $y - 2 = -3(x + 1)$ or $y + 4 = -3(x - 1)$
3. $y - 2 = \frac{1}{3}(x - 3)$ or $y = \frac{1}{3}(x + 3)$

4. $y - 5 = 3(x - 2)$ **5.** $y - 4 = 2(x - 1)$
6. $y = \frac{1}{2}(x + 2)$ **7.** $y - 7 = x - 3$
8. $y - 8 = -4(x + 5)$ **9.** $y + 4 = 9x$
10. $y = 1$ **11.** $y + 4 = -2(x + 3)$
12. $y + 10 = 5(x - 6)$ **13.** $y = 5x$
14. $y - 3 = 2(x - 2)$ **15.** $y - 6 = 3(x - 9)$
16. $y + 7 = -(x - 8)$ **17.** $y + 2 = 7(x - 1)$
18. $y + 7 = \frac{1}{2}(x - 2)$ **19.** $y + 4 = 3(x + 5)$

20. $y - \frac{1}{3} = -7\left(x + \frac{1}{2}\right)$ **21.** $y + 8 = \frac{2}{3}(x - 4)$

22. $y - 3 = 2(x - 1)$ or $y - 9 = 2(x - 4)$

23. \$13 **24.** $d = \frac{13}{14}t$ **25.** about 6:42 P.M.

Practice B

1. $y + 2 = 2(x + 1)$ or $y - 2 = 2(x - 1)$

2. $y - 1 = \frac{2}{5}(x - 2)$ or $y + 1 = \frac{2}{5}(x + 3)$

3. $y + 1 = -\frac{1}{2}(x + 4)$ or $y + 4 = -\frac{1}{2}(x - 2)$

4. $y - 24 = -2(x + 3)$
5. $y + 2 = -5(x + 4)$ **6.** $y + 3 = \frac{2}{3}x$
7. $y + 5 = -4(x - 6)$ **8.** $y = 6$
9. $y + 5 = 6(x + 3)$ **10.** $y - 1 = -6(x + 12)$
11. $y - 21 = -\frac{1}{3}(x + 14)$
12. $y - 4 = -\frac{2}{3}(x - 16)$ **13.** $y = x + 5$
14. $y = 4x - 21$ **15.** $y = 2x - \frac{25}{2}$ **16.** $y = 5x + 6$
17. $y = -2x + 1$ **18.** $y = 3x - 7$
19. $y = -3x + 16$ **20.** $y = \frac{1}{2}x - 12$
21. $y = 4x + \frac{7}{3}$
22. $y - 3.5 = 1.5(x - 1)$ or $y - 8 = 1.5(x - 4)$
23. \$11 **24.** $y = 7x$ **25.** about 9:46 A.M.

Reteaching with Practice
1. $y = 2x - 3$ **2.** $y = -3x + 3$
3. $y = -4x$ **4.** $y = -4x + 7$ **5.** $y = -x - 8$

Interdisciplinary Application
1. $y - 45 = 10(x - 3)$

2.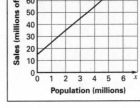

3. 85 million pesos

4. $y - 15 = 7.3(x - 3)$

5.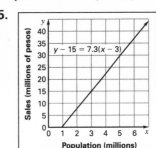

6. 36.9 million pesos

Lesson 5.2 *continued*

Challenge: Skills and Applications

1. $y - 11 = 10(x - 2)$ or $y + 4 = 10\left(x - \frac{1}{2}\right)$

2. $y - 3 = \frac{3}{5}(x - 8)$ or $y + 2 = \frac{3}{5}\left(x + \frac{1}{3}\right)$

3. $y - 0.9 = \frac{1}{2}(x + 0.5)$ or $y + 0.5 = \frac{1}{2}(x + 3.3)$

4. $y + 1.4 = -4(x - 3.2)$ or
   $y - 1.8 = -4(x - 2.4)$

5. $y - 4 = -2\left(x + \frac{2}{3}\right)$ or $y + \frac{1}{3} = -2\left(x - \frac{3}{2}\right)$

6. $y + 4 = -\frac{5}{6}(x - 5)$ or $y + \frac{1}{4} = -\frac{5}{6}\left(x - \frac{1}{2}\right)$

7. $y - q = -\frac{q}{2p}(x - p)$ or $y - 2q = -\frac{q}{2p}(x + p)$

8. $y + q = -(x - 2p)$ or $y + q - p = -(x - p)$

9. $y - 3 = -\frac{5}{2}(x - 6)$

10. $(4, 8)$; *Sample explanations*: Method 1: Substitute $2x$ for y in the equation from Exercise 9 and then solve for x; Method 2: Find points on the line by using the slope $-\frac{5}{2}$ to count up or down from $(6, 3)$. As x increases, y decreases so y will not become twice x. Therefore, you must decrease x and increase y. Keep subtracting 1 from x and adding $2\frac{1}{2}$ to y until you get to the point $(4, 8)$ that works.

11. $(12, -12)$ 12. $(2, 13)$

13. $y - 5620 = 280(x - 1)$ or
    $y - 6040 = 280(x - 2.5)$ 14. 6740 ft 15. 4780 ft

Quiz 1

1. $y = -3x + 7$ 2. $y = -\frac{3}{5}x - 3$

3. $y = 4x + 3$ 4. $y = \frac{2}{3}(x - 1)$

5. $y = 2x + 10$ 6. $y = \frac{1}{4}x + 5$

7. $y = -5x + 14$

Lesson 5.3

Warm-Up Exercises

1. $y = \frac{1}{2}x + 6$ 2. $y = -2x + 9$

3. $y = -3x + 7$ 4. $y = \frac{1}{4}x + \frac{1}{2}$

Daily Homework Quiz

1. $y - 4 = \frac{3}{4}(x - 1)$

2. $y - 5 = -3(x + 1)$

3. $y = -4x - 2$

4. $y = \frac{1}{4}x - \frac{1}{2}$

5. $7.35 per hour

Lesson Opener

Allow 10 minutes.

1. $(5, 30), (3, 18)$ 2.

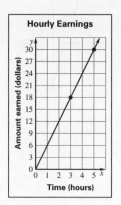

3. 6   4. 0   5. $(2, 14), (5, 20)$

6.

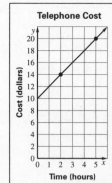

7. 2   8. 10

9. Yes; you know the slope and the y-intercept so you can use the slope-intercept form of the line to find the equation.

Practice A

1. $y = x + 3$ 2. $y = x$ 3. $y = -2x + 1$

4. $y = \frac{1}{4}x - 2$ 5. $y = 2x - 1$

6. $y = -3x - 4$ 7. $y = -2x$ 8. $y = x + 4$

9. $y = -3x - 6$ 10. $y = x + 2$

11. $y = \frac{1}{3}x - \frac{5}{3}$ 12. $y = x$ 13. $y = -4x - 3$

14. $y = 3x + 16$ 15. $y = 2x - 1$ 16. $y = 4x$

17. $y = \frac{18}{7}t + 51$ 18. $y = -\frac{1}{3}x + \frac{8}{3}$

19. $y = \frac{11}{15}x + 50; \frac{11}{15}$

Practice B

1. $y = x - 3$ 2. $y = -3x - 5$

3. $y = 4x - 3$ 4. $y = -\frac{1}{3}x + 2$

5. $y = 2x - 1$ 6. $y = -\frac{3}{2}x + 3$

7. $y = 5x + 8$ 8. $y = -6x - 33$

Lesson 5.3 *continued*

9. $y = \frac{3}{4}x - \frac{1}{4}$ **10.** $y = -3x + 14$
11. $y = 2x + 4$ **12.** $y = 5x + 31$
13. $y = x - \frac{3}{2}$ **14.** $y = -0.5x - 0.64$
15. $y = -\frac{17}{5}x + \frac{13}{10}$
16. $y = -\frac{2}{5}x + \frac{16}{5}; y = -\frac{2}{5}x - \frac{13}{5};$
the slopes are the same.   **17.** $y = 60t - 20$

Reteaching with Practice

1. $y = x + 5$ **2.** $y = -8x + 7$ **3.** $y = 3x + 3$
4. $y = -2x + 4$ **5.** $y = \frac{7}{3}x + 12$

Interdisciplinary Application

1. $y = 393x + 14324$

2.
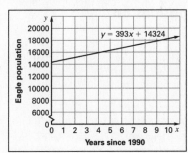

3. 18,254   **4.** $y = 518.25x + 13,697.75$

5.
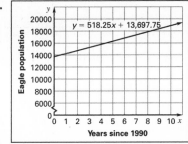

6. about 18,880

Challenge: Skills and Applications

1. $y = -\frac{20}{9}x$ **2.** $y = 2x + 1$
3. $y = -\frac{1}{3}x - 1$ **4.** $y = 6x - 17$
5. $y = -\frac{1}{6}x + \frac{3}{2}$ **6.** $y = 6x + 20$
7. $y = -\frac{1}{6}x + \frac{23}{3}$ **8.** $y = 0.625x - 2.75$, or
$y = \frac{5}{8}x - \frac{11}{4}$ **9.** about 24 days   **10.** 9.75 in.
11. -2.75 inches; *Sample answers*: One
possibility is that the seed is planted $2\frac{3}{4}$ in. below
ground so it starts off with a height of $-2\frac{3}{4}$ in.
Another possibility is that the linear growth model

in Exercise 8 is not appropriate until the plant has
reached a certain stage of growth. Extending the
line to the left results in a negative value for $x = 0$,
but the values obtained from the model may not
reflect the real-life situation for the first few days.

Lesson 5.4

Warm-Up Exercises

1. $y - 3 = -2(x + 1)$
2. $y + 5 = -(x - 3)$ or $y + 4 = -(x - 2)$
3. $y = -2x + 9$ **4.** $y = \frac{1}{2}x + 6$

Daily Homework Quiz

1. $y - 3 = \frac{1}{3}(x - 4)$ or $y - 2 = \frac{1}{3}(x - 1)$
2. $y = -2x + 4$
3. $y = 2x - 3$
4. $y = 3x$

Lesson Opener

Allow 15 minutes.

1, 2. The equations should be matched as follows:
$y = 2x + 1, -2x + y = 1;$
$y = 4x - 1, 4x - y = 1;$
$y = \frac{1}{5}x + 2, -x + 5y = 10;$
$y = -x + 3, x + y = 3;$
$y = -\frac{1}{2}x - 4, x + 2y = -8;$
$y = 5x + 1, 5x - y = -1;$
$y = x - 6, x - y = 6;$
$y = \frac{2}{3}x + 3, -2x + 3y = 9$

Practice A

1. $x - y = 9$ **2.** $6x - 4y = -7$ **3.** $x = -7$
4. $2x + 3y = 6$ **5.** $11x + y = -4$ **6.** $y = 1$
7. $4x - y = -3$ **8.** $x - 8y = -2$
9. $x - y = 0$ **10.** $5x - y = \frac{1}{2}$
11. $x - 4y = -12$ **12.** $2x + 3y = -3$
13. $x - y = -4$ **14.** $3x + y = 11$
15. $-8x + y = 11$ **16.** $4x - y = 31$
17. $2x + y = 16$ **18.** $2x + y = -17$
19. $x - y = 7$ **20.** $5x - y = -3$
21. $2x + y = -4$ **22.** $-6x + y = 1$
23. $3x + y = -9$ **24.** $-4x + y = 10$
25. $2x + y = 30$

Lesson 5.4 *continued*

Answers

26. 30, 20, 14, 10, 0

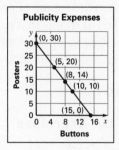

Publicity Expenses

Posters / Buttons

27. $10x + 12y = 240$

28. 20, 15, 10, 5, 0

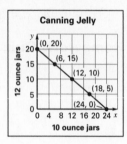

Canning Jelly

12 ounce jars / 10 ounce jars

Practice B

1. $2x - y = 8$ **2.** $3x - 4y = 75$
3. $-3x + y = 2$ **4.** $3x + y = 5$
5. $6x - 21y = 18$ **6.** $2x - 3y = 5$ **7.** $x = 4$
8. $y = 4$ **9.** $10x - 3y = 45$
10. $x - 8y = -12$ **11.** $-x + 2y = 8$
12. $-2x + 3y = -5$ **13.** $-2x + y = -5$
14. $4x + y = 9$ **15.** $-3x + y = 6$
16. $6x + y = -8$ **17.** $-x + 3y = -30$
18. $x + 2y = 6$ **19.** $-3x + y = -7$
20. $3x + y = -1$ **21.** $x + 8y = 17$
22. $-5x + y = 15$ **23.** $-x + 2y = -6$
24. $x + 3y = 12$ **25.** $y = -4, x = 3$
26. $y = 1, x = 5$ **27.** $y = -2, x = -3$
28. $y = -4, x = 0$ **29.** $2x + 3y = 60$
30. $y = -\frac{2}{3}x + 20$

31.

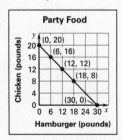

Party Food

Chicken (pounds) / Hamburger (pounds)

32. 20, 16, 12, 8, 0

33. $4x + 6y = 48$ **34.** $y = -\frac{2}{3}x + 8$

35.

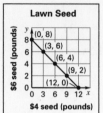

Lawn Seed

$6 seed (pounds) / $4 seed (pounds)

36. 8, 6, 4, 2, 0

Reteaching with Practice

1. $2x - 3y = 21$ **2.** $2x - y = -8$
3. $x + 4y = 24$ **4.** $2x + y = 6$
5. $3x - y = -10$ **6.** $x + y = 3$
7. $5x + y = 8$

Real-Life Application

1. $8x + 10y = 200$

2.

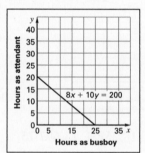

Hours as attendant / Hours as busboy / $8x + 10y = 200$

3. $1440

4. $3b + 2c = 300$ **5.**

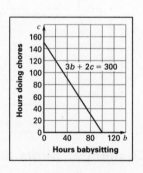

Hours doing chores / Hours babysitting / $3b + 2c = 300$

Challenge: Skills and Applications

1–4. Accept equivalent equations in standard form. **1.** $74x - 12y = 45$ **2.** $5x - 4y = -24$
3. $(p + 4)x + 3y = 2p - 4$
4. $(q + 2)x + 2y = 3q + 2$ **5.** $2x + 3y = 12$
6. 2.8 ml **7.** yes; 2.4 ml
8. $4.5x + 3.75y = 24$; 2 that are 4.5 ft and 4 that are 3.75 ft **9.** no; if he buys two 12-foot lengths, he can cut 2 shelves that are 3.75 ft and 1 shelf that is 4.5 ft from each

A6 **Algebra 1**
Chapter 5 Resource Book

Lesson 5.4 *continued*

Quiz 2

1. $y = -2x + 10$ 2. $y = 5x + 3$
3. $3x - 5y = 25$ 4. $x - 2y = -12$
5. $x - 2y = -10$

Lesson 5.5

Warm-Up Exercises

1. $y = 2x + 6$ 2. $y = 2x + 5$
3. $y = 2x + 4$ 4. $y = 2x + 3$

Daily Homework Quiz

1. $12 - 4y = -1$ 2. $2x + y = -2$
3. $-2x + y = 12$ 4. $y = -3$
5.

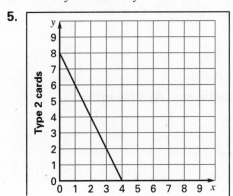

Lesson Opener

Allow 10 minutes.

1. $-x + y = 1$ 2. $5x + 6y = 4$
3. $-5x + 4y = 12$ 4. $y = 2$
5. $y = -\frac{1}{2}x$ 6. $-2x + y = -1$

Practice A

1. 12 2. 8 3. $y = 12x + 8$ 4. $80 5. 3 days
6. 55 7. $246 8. $576 9. 4 10. $5.50
11. $7.00 12. $4x + 2y = 12$
13. Values in the table are: 6, 4, 2, 0
14. As the number of tomato seedlings you can purchase increases, the number of pepper seedlings decreases.

Practice B

1. 75 2. 50 3. $y = 75x + 50$ 4. $275
5. 5 hours 6. 4.2 7. $36,500,000
8. $49,100,000 9. 35 10. $36,800
11. $38,000 12. $3x + 1.5y = 9$
13. Values in table are: 6, 4, 2, 0
14. As the number of pounds of grapes you can buy increases, the number of pounds of plums decreases.

Reteaching with Practice

1. $y = 3t$ 2. $y = 100t + 500$
3. 6 miles 4. $1100
5. $40x + 20y = 120$;

Jeans, x	0	1	2	3
Shirts, y	6	4	2	0

Learning Activity

1–2. Check answers.

Real-Life Application

1. 10,157 2. 29,682
3. The model from Exercise 2 because it is based on the most recent data. 4. 97,392

Challenge: Skills and Applications

1. $y = 3x - 4$ 2. $y = -2x + 8.33$
3. $y = -10.56t + 358$
4. The exchange rate is decreasing; the yen is getting stronger.
5. $y = -10.4t + 356.7$
6. Estimate based on Ex. 3 model: 41.2; Estimate based on Ex. 5 model: 44.7

Lesson 5.6

Warm-Up Exercises

1. $-2x + y = 4$ 2. $4x + 3y = -12$
3. $-x + y = -7$ 4. $-7x + y = -23$
5. $-x + 4y = 6$ 6. $-x + y = -5$

Answers

Lesson 5.6 *continued*

Daily Homework Quiz

1. 7 ft/sec
2. 0 ft
3. $y = 7t$
4. no   5. 70 sec

Lesson Opener

1. $m_l = 2, m_m = -\frac{1}{2}$ 2. -1
3. $-\frac{1}{3}$ 4. -5

Graphing Calculator Activity

1. a. -3 b. $\frac{1}{2}$ c. $-\frac{2}{3}$ 2. *Sample answer:*
$y = \frac{1}{2}x + 5, y = -2x, y = -2x + 5$

Practice A

1. yes   2. no   3. yes   4. yes
5. no   6. yes   7. yes   8. yes
9. no   10. yes
11. $y = -\frac{1}{2}x + 4$ 12. $y = x + 11$
13. $y = -4x + 14$ 14. $y = -\frac{1}{2}x + 1$
15. $y = -\frac{1}{2}x + 3$ 16. $y = \frac{8}{3}x - 2$
17. $y = -\frac{3}{2}x + 5$ 18. $y = \frac{1}{5}x - 7$
19. $l_1: y = \frac{4}{3}x + 4, l_2: y = -\frac{1}{2}x + 1$
$\left(\frac{4}{3}\right)\left(-\frac{1}{2}\right) \neq -1$, not perpendicular
20. $l_1: y = 4x - 4, l_2: y = -\frac{1}{4}x - 1$
$(4)\left(-\frac{1}{4}\right) = -1$, perpendicular
21. $l_1: y = x; l_2: y = -x$
$(1)(-1) = -1$, perpendicular

Practice B

1. yes   2. yes   3. no   4. no   5. no   6. yes
7. $y = 2x + 12$ 8. $y = -3x + 13$
9. $y = -3x + 20$ 10. $y = -2x - 4$
11. $y = -\frac{3}{2}x + 6$ 12. $y = -2x - 3$

13. $y = -\frac{1}{6}x - 1$ 14. $y = -x - 1$
15. $y = \frac{1}{5}x + 10$ 16. $y = \frac{4}{3}x + \frac{1}{2}$
17. $y = -\frac{3}{5}x + \frac{3}{4}$ 18. $y = \frac{2}{3}x - 8$
19. $y = -x + 6$ 20. $y = \frac{5}{4}x - \frac{11}{2}$
21. $y = \frac{3}{7}x + \frac{48}{7}$ 22. $y = -\frac{8}{3}x - \frac{38}{3}$

Reteaching with Practice

1. yes   2. no
3. $y = -\frac{1}{3}x - \frac{8}{3}$; $(3)\left(-\frac{1}{3}\right) = -1$
4. $y = -\frac{3}{2}x$; $\left(\frac{2}{3}\right)\left(-\frac{3}{2}\right) = -1$
5. $y = -\frac{1}{2}x + 5$
6. $y = 3x - 14$
7. $y = \frac{1}{4}x + 2$

Interdisciplinary Application

1. pentagon

Challenge

1. yes   2. yes   3. no   4. no
5. yes   6. yes   7. $y = -4x + 6$
8. $y = 0.3125x - \frac{1}{2}$ 9. $y = 0.625x$
10. $y = -5x - 8$
11. $(2, -5)$ 12. $(-6, -3)$
13. $(-4, -1)$ 14. $(-3, -3)$
15. Slope $\overline{ZW} = -\frac{2}{5}$, slope $\overline{ZY} = \frac{5}{2}$; $\overline{ZY}$ and $\overline{ZW}$ are perpendicular since $\left(-\frac{2}{5}\right)\left(\frac{5}{2}\right) = -1$.
Slope $\overline{YX} = -\frac{2}{5}$, slope $\overline{ZY} = \frac{5}{2}$; $\overline{YX}$ and $\overline{ZY}$ are perpendicular since $\left(-\frac{2}{5}\right)\left(\frac{5}{2}\right) = -1$.
16. line through $\overline{ZW}$: $y = -\frac{2}{5}x + \frac{16}{5}$; line through $\overline{ZY}$: $y = \frac{5}{2}x + 9$; line through $\overline{YX}$: $y = -\frac{2}{5}x - \frac{13}{5}$
17. line through $\overline{ZW}$: $y = -\frac{2}{5}x + \frac{16}{5}$; line through $\overline{YX}$: $y = -\frac{2}{5}x - \frac{13}{5}$; $\overline{ZW}$ and $\overline{YX}$ are parallel since their slopes are equal.

Review and Assessment

Review Games and Activities

1. $6x + 5y = -10$ **2.** $y = \frac{1}{2}x + 3$

3. $3x - 7y = 49$ **4.** $8x - 7y = 70$

5. $y = -\frac{13}{6}x - \frac{23}{3}$ **6.** $6x - 7y = -42$

7. $y = -\frac{5}{8}x + \frac{21}{8}$ **8.** $2x - y = 3$

9. $y = \frac{1}{2}x - 1$ **10.** $x + 3y = 9$

POPLAR ONES

Chapter Test A

1. $y = -5x + 7$ **2.** $y = 10x - 3$

3. $y = x + 1$ **4.** $y = 5x + 5$

5. $y = -7x + 70$ **6.** $y = -2x + 6$

7. $y = 2x$ **8.** $y = \frac{2}{3}x + \frac{7}{3}$ **9.** $y = \frac{3}{2}x - 2$

10. $y = x - 5$ **11.** $y = 2x + 9$

12. $y = -\frac{3}{5}x - \frac{2}{5}$ **13.** $y = \frac{6}{5}x + \frac{7}{5}$

14. $y = -\frac{1}{2}x + \frac{11}{2}$ **15.** $y + 4 = \frac{4}{3}(x + 3)$ or

$y - 4 = \frac{4}{3}(x - 3)$ **16.** $y + 4 = -\frac{1}{12}(x + 5)$ or

$y + 5 = -\frac{1}{12}(x - 7)$ **17.** $5x - y = -6$

18. $3x + y = 9$ **19.** $y = 3, x = 3$

20. $y = 2, x = -1$ **21.** $y = x - 3$

22. $y = x + 2$

Chapter Test B

1. $y = -3x + 5$ **2.** $y = 4x$ **3.** $y = \frac{3}{4}x + 3$

4. $y = \frac{2}{3}x - 2$ **5.** $y = 3x + 4$ **6.** $y = \frac{1}{2}x + \frac{1}{2}$

7. $y = \frac{1}{2}x + \frac{7}{2}$ **8.** $y = \frac{2}{7}x + \frac{13}{7}$

9. $y = -\frac{6}{5}x + \frac{8}{5}$ **10.** $y = -3x + 9$

11. $y = \frac{1}{2}x + \frac{1}{2}$ **12.** $y = -\frac{8}{9}x - \frac{13}{9}$

13. $y = x - \frac{1}{2}$ **14.** $y = \frac{1}{3}x + \frac{5}{3}$

15. $y + 6 = \frac{1}{4}(x - 5)$ or $y + 7 = \frac{1}{4}(x - 1)$

16. $y + 3 = -\frac{12}{7}(x - 6)$ or

$y - 9 = -\frac{12}{7}(x + 1)$ **17.** $2x - 8y = 3$

18. $x + 3y = -15$ **19.** $y = 4, x = 2$

20. $y = 4, x = -5$ **21.** $10x + 15y = 55$

22. $y = x$ **23.** $x = 0$

SAT/ACT Chapter Test

1. B **2.** C **3.** D **4.** C **5.** B **6.** D **7.** A

8. C **9.** A

Alternative Assessment

1. a–d, Complete answers should address these points. • Explain that linear models are linear functions that are used to model a real-life situation. The rate of change is a method of comparing the change of two quantities.

a. The situation should be able to be modeled with a linear equation. It should involve a rate of change. At least two data points should be given.
b. A linear equation based on the points chosen in part (a) should be given in slope-intercept or point-slope form. **c.** Check graphs. The graph should be linear. **d.** The point should fall on the line or the student should have explained why the prediction was not accurate.

2. a. 75; -7.5; Jeremy deposits $75 weekly; Zachary withdraws $7.50 weekly. **b.** Jeremy: $y = 75x + 275$; Zachary: $y = -7.50x + 522.50$ **c.** Check graphs. **d.** Jeremy: $1775; Zachary: $372.50 **e.** 70 weeks; He can withdraw $5 in the last week.

3. *Critical Thinking Solution* They would be the same after 3 weeks. *Sample answer:* One method would be to set the two equations equal to each other, $75x + 275 = -7.50x + 522.50$, and solve for x. Another method would be to sketch the graph of both equations and find out where the two equations have the same x-value. This would occur where the two equations intersect. A third method would be to use a table of values, either on the calculator or by hand, and calculate the balances for several weeks to determine the solution.

Project: What We Read

1. $y = 0.95x + 17.5$; $y = 0.45x + 23.8$
2. $20.35 billion; $25.15 billion
3. The estimate for books and maps was closer.
4. $27 billion; $28.3 billion
5. Answers will depend on data in 2000.

Answers

Review and Assessment *continued*

Cumulative Review

1. -343 **2.** $\frac{13}{27}$ **3.** -0.981 **4.** -217

5. $l = 3w$ **6.** $t < \frac{1}{2}s$ **7.** $h = \frac{2a}{b}$ **8.** -3

9. 21.8 **10.** 2.2 **11.** $-\frac{49}{18}$ **12.** -7

13. 8 **14.** 33.77 **15.** 4

16. x-intercept $= \frac{7}{2}$, y-intercept $= -14$

17. x-intercept $= -5$, y-intercept $= 4$

18. -12 **19.** $\frac{1}{21}$ **20.** $y = 3x - 2$

21. $y = -5x + 4$ **22.** $y = 7x + 11$

23. $y = \frac{1}{5}x - 6$ **24.** $y = 8$

25. $y = -6.5x + 4.5$ **26.** $y = -\frac{1}{5}x - \frac{3}{5}$

27. $y = -2x - 80$ **28.** $y = -3x + 8$

29. $y = -x$

30. $y - 7 = \frac{7}{2}(x - 2)$ or $y + 7 = \frac{7}{2}(x + 2)$

31. $y - 8 = 2(x + 10)$ or $y + 12 = 2(x + 20)$

32. $y - 1 = \frac{2}{3}(x - 4)$ or $y + 3 = \frac{2}{3}(x + 2)$

33. $y - 10 = -4(x + 10)$ or $y + 6 = -4(x + 6)$

34. $y = 4x - 18$ **35.** $y = -\frac{1}{4}x + 2$

36. $y = -3x + 7$ **37.** $y = \frac{4}{3}x - 3$

38. $5x + y = 5$ **39.** $x + 2y = 10$

40. $12x + 2y = -9$ **41.** $4x - y = -20$

42. $y = -0.442x + 24.3$ **43.** about $11,000,000$

44. about $8,800,000$ **45.** $y = -x$

46. $y = \frac{1}{3}x + \frac{7}{3}$ **47.** $y = \frac{4}{3}x - \frac{11}{3}$